Do-It-Yourself High-Performance Car Mods

Rule the Streets

Matt Cramer

Mc
Graw
Hill
Education

New York   Chicago   San Francisco
Lisbon   London   Madrid   Mexico City
Milan   New Delhi   San Juan
Seoul   Singapore   Sydney   Toronto

Do-It-Yourself High-Performance Car Mods: Rule the Streets

1 2 3 4 5 6 7 8 9 0   DOC/DOC   1 9 8 7 6 5 4 3

ISBN   978-0-07-180409-7
MHID   0-07-180409-9

This book is printed on acid-free paper.

Sponsoring Editor Judy Bass	**Copy Editor** Lisa McCoy
Editing Supervisor Stephen M. Smith	**Proofreader** Paul Tyler
Production Supervisor Richard C. Ruzycka	**Indexer** Jack Lewis
Acquisitions Coordinator Bridget L. Thoreson	**Art Director, Cover** Jeff Weeks
Project Manager Patricia Wallenburg, TypeWriting	**Composition** TypeWriting

Contents

Preface

Spend time on almost any Internet forum dedicated to cars and you'll see a steady procession of beginning car enthusiasts who aren't sure where to start. There's a lot to learn, and a few 200-word messages aren't enough to really explain just how everything in a car works. Having spent a lot of time on Internet forums, there have been many times I've wanted to tell a beginner, "I think what you really need is a good book."

There are many excellent books if you want a very in-depth discussion of a particular part of a car, such as the engine, suspension, or electronics, and how to improve it. I've listed several of these at the end of this book if you want to learn more. There are also several good books for understanding the parts of a car if you want to fix one to work like new. However, there hasn't been much in the way of books that provide a basic working knowledge of the whole car with an eye to modifying it to go faster. When looking for the best single book to recommend to someone who's just getting started, I couldn't really find such a thing. So I decided to write it myself.

This book is meant to be a starting point. While I can't give in-depth answers on everything, I can give you the right questions. If you're looking through a catalog trying to pick the best springs for your Subaru, I can't tell you which ones on the market will be best for you (especially since there are probably going to be a couple of brands that make it to market between the time I write this sentence and the time you read it, and some that have been discontinued too). But this book will explain how to evaluate how your Subaru drives and how you can figure out if you need to stiffen the front or the rear springs—or maybe how you need to skip the springs at this point because what you really need is a good set of struts.

There's a cliché about life being a journey instead of a destination. While not original, the metaphor is a hard one to resist for a book about cars. If you're taking a journey into the world of modifying cars, I can show you the on-ramp and how to get to some of my favorite places. I hope you'll start from there and go discover more, find where you like to visit, and discover new roads. Happy trails to you!

Matt Cramer

Acknowledgments

We all have to learn somewhere, and it can give you a big head start to learn from others' experience instead of having to experience it all for yourself. Several people I've learned from who I specifically want to thank, whether they taught me directly or I learned from things they wrote, include David Vizard, Fred Puhn, Hugh MacInnes, Corky Bell, Jeff Hartman, A. Graham Bell (not the one who invented the telephone), Chris Macellaro, Scott Clark, and Daniel Stern. A lot of those guys also have books or websites that I'd recommend seeking out after you've finished this book. I'd also like to thank Jerry Hoffmann, who I collaborated with on a previous book, and who gave me a job working in the racing industry full time. Actually, thanks for that are probably in order for the MegaSquirt team as well, including Bruce Bowling, Al Grippo, Lance Gardiner, James Murray, Ken Culver, Jean Belanger, Phil Ringwood, Phil Tobin, Eric Law, and many others who made various contributions, big and small.

I'd also like to thank the representatives of several performance companies who helped me out with photos their companies had on file. A lot of the illustrations here would have been a lot cruder if they hadn't already had professional artwork ready to go. I've credited the sources throughout the text, although the pictures are often work for hire, and I frequently don't have the original artists' names, much as I'd like to give them all personal thank yous too.

Professional racing teams are never a one-person operation, and neither is professional book publishing. There are several others I would like to thank personally, including the acquisitions editor, Judy Bass, and Patricia Wallenburg, who put in plenty of hours making sure the text not only had correct grammar, but made sense. And there are several others working behind the scenes whose names I don't know, but they also deserve credit for making the book look like a finished product and not just a big wall of text. I would also be remiss if I did not mention Michael Lutfy, who was the editor on my first book and encouraged me to get into writing.

And to my wife, Kelly, a big thank you for putting up with all the time I've been spending at the computer typing this thing out.

Unit Conversions

Car parts may be measured in a variety of units. Sometimes you may need to convert one unit to another. This chart lists some of the more commonly used units encountered while working on cars.

To convert from these units . . .	. . . to these units . . .	. . . multiply by this number.
Millimeters	Inches	0.03937
Inches	Millimeters	25.4
Meters	Feet	3.281
Feet	Meters	0.3048
Liters	Cubic inches	61
Cubic inches	Liters	0.0164
Liters	Cubic centimeters	1000
Cubic centimeters	Liters	1/1000
cc/minute	Gallons per hour	0.01585
Gallons per hour	cc/minute	63.1
Gallons per hour	Pounds per hour*	6.0
Pounds per hour*	Gallons per hour	1/6
cc/minute	Pounds per hour	0.0951
Pounds per hour	cc/minute	10.52
Cubic feet per minute	Liters per minute	28.31
Liters per minute	Cubic feet per minute	0.035315

Pounds per hour applies to gasoline only.

About the Author

Matt Cramer is an engineer and automotive specialist for the race car electronics firm DIYAutoTune.com—and has a wealth of experience teaching customers how to understand, use, install, and troubleshoot electronics and other parts. Also a freelance journalist for print and online media, he has written a monthly column for *Miata Journal* along with articles that have appeared in *Grassroots*, *Motorsports*, and *Jalopnik*. Mr. Cramer resides in Covington, Georgia.

CHAPTER 1

Planning

Getting a Clearer Vision of the Car of Your Dreams

Remember that fable about the tortoise and the hare? Well, it's not just a fable; it's close to the way a lot of racing events play out. You don't see people lose too often for falling asleep on the course, but there are a lot of racers who try to go fast but fail to keep up a steady pace for other reasons. I can picture Aesop hanging around the chariot races and laughing at the charioteer who had built a set of super-light wheels only to find they weren't able to hold together for a whole race, or the racer who had put together a team of the swiftest horses only to find that the beasts refused to work together as a team. Meanwhile, the guy who didn't have any fancy parts, but

A great project car isn't just a result of great parts—the parts need to be chosen to work well together. This MR2 doesn't use many super-high-dollar parts so much as an effective combination of parts chosen to fit a reasonable budget. (*Photo courtesy DIYAutoTune.com.*)

had spent a lot of time just practicing and making sure everything he had worked together went on to take the win. At amateur car races today, many cars that seem like they should be faster lose either because the car can't hold together for a full race, or because the owner spent more time searching for exotic parts than tuning the car to get everything working well together. Well-sorted and steady wins the race. Planning a project car well is the best way to end up in the winners' circle.

Although I've seen some people find a rare or exotic part and then start thinking of what sort of car to put it on, normally you won't want to start your planning with what parts to use. The best place to start planning is with three decisions: what car, what budget, and what purpose. You might already have a particular car and want to make it the fastest example of it in town. Or you might want to compete in a particular event, make sure you can afford it, and then go shopping for an example of the car you expect to do best.

Picking these three wisely can make or break a project. Picking a goal that your choice of car isn't good at can drive the budget up unnecessarily. Some goals pretty much come with their budget handed to you along with them—you're going to have a difficult time competing in Grand Am racing on less than a six-figure budget. If you ever plan on selling the car, it's even possible to set too big a budget, resulting in a car that nobody else is likely to be interested in buying when you're done with it. The most common mistake I've seen many beginners make, however, is losing track of their goals.

Lee Sicilio's Dodge Charger was built with a single goal in mind: to push a 40-year-old car with minimal aerodynamic modifications past 300 mph. (*Photo courtesy DIYAutoTune.com.*)

Keeping your goals in focus is a point I'd like to emphasize here. Modifying a car frequently means making trade-offs. Often, to get more of one goal, you are likely to compromise your car's other attributes. For example, someone who wants less weight may remove the air conditioner, trading comfort for speed. More horsepower often means burning more gasoline, bringing down gas mileage. In many cases, you will find that the trade-offs are between performance and practicality, or between performance and cost. You will need to decide which characteristics of your car you want to improve, and which are less important.

Many beginners want to build their car into an automotive pentathlete that can accelerate better, handle better, wow everyone at car shows, still be a practical commuter car, and maybe even have a really serious stereo system. While it is possible to build a car that is a jack of all trades, as the cliché goes, the car would be a master of none. While you can give your car better handling, distinctive looks, louder speakers, and more power, such a car will often be less comfortable and less reliable for your daily commute. Furthermore, such a car would not be able to compete with a purpose-built car built on the same budget. The "jack of all trades, master of none" saying has a little-known second half—"though often better than a master of one." This doesn't really apply to cars; the cars that dominate racing events are nearly always examples of the "master of one." Of course, if you just want to have a bit more fun with your street car and aren't out to win, this may not be what you want.

Performance goals themselves tend to become incompatible at the "master of one" level. Adding just a little more power to the engine may not hurt handling, but a heavily built engine may produce so much power that it smokes the tires any time you put the throttle down, or might deliver power in such an unpredictable way that it is hard to drive smoothly. A suspension built to handle is likely to cost the car traction off the line,

This kit car, based on a Lotus Seven design, uses its light weight and carefully tuned suspension to make it a very difficult car to beat on a tight autocross course.

Building a car that can perform in several different events is not easy. Andrew Nelson's Fiat uses different suspension setups for handling and drag racing.

while a suspension designed entirely for drag racing will seldom handle well outside of a straight line. Likewise, show modifications tend to add weight, while racing a show car means constantly having to clean off rubber and repair rock chips, or worse, repairing the aftermath of hitting another race car.

When deciding on your priorities, consider how they fit your choice of car. It is often easier to ruin a car's strengths than to fix its weaknesses, particularly if you just start throwing mods at it without careful planning. If you are starting with a Lexus ES330, for example, this car is designed for an incredibly soft, quiet ride. It is much easier to turn the Lexus into a noisy car with a harsh ride than it is to turn it into a world-class handling machine. A carefully planned project might give the ES330 world-class cornering ability, but it would still lose its feeling of quiet luxury.

Looking for a Car

Shopping for a project car can be a very different experience from looking to buy basic transportation. The car's strengths should match your goals, unless you want the extra challenge of making a car do something that goes against conventional wisdom. For example, building a V8-powered Chevy Nova for brutally fast acceleration is relatively straightforward. Making it handle somewhat better than stock is also reasonable. Trying to make the Nova handle like a race-prepared Miata or to make a four-cylinder early Nova run fast without an engine swap would be extremely difficult.

Its small size and excellent aftermarket support make a Miata an excellent choice for a car built to handle.

Whether you want wildly customized bodywork, brutal horsepower, or snappy handling, make sure the car's capabilities and the aftermarket parts you can find match your goals. You may have to research the parts available for the car you are considering to be sure you can find the right mods to make it what you want it to be. Although you can get custom-made performance parts for nearly anything if you have enough determination and money, it is often much cheaper and considerably less hassle if you can find a company that is already mass producing what you need. Sometimes, you may discover that there simply are not any parts available off the shelf for its engine or suspension. It may be frustrating to give up on a particular model in the planning stages if you've fallen in love with a particular car, but it can be even more frustrating to buy the car and discover that there is no way you can afford to make it reach your goal.

You will also need to decide whether you want the car to be your main means of transportation or a toy to drive on weekends. If you have the money and storage space, having two cars protects you from having to wonder how you will get to work in the morning if a part installation takes longer than planned. A second car is not just for the rich, as a car with mod potential can be found for as little as you want to pay. Domestic V8 lovers often have the option of finding an engineless car for little more than scrap value and dropping in a motor from the junkyard. Sport compact fans can find a selection of older turbocharged coupes or featherlight handling machines on a shoestring budget with enough looking. More mechanically inclined types may even find it viable to mod their daily driver and keep a dirt-cheap car or pickup for when their better car is not back together. Working on a second car can be a viable option for those who have room to keep a project car, do not need to have the latest model, and are willing to put in the time and effort needed to clean up a cheap car or the money to pay for two cars already in good shape.

The performance parts industry doesn't have much love for the Chevy LUV trucks. The only way one of these would be a good choice for drag racing is if you are already planning to swap engines. It took a lot of custom fabrication to get the truck to this point.

Project cars are out there for every budget. This is one of the cars that competed in Grassroots Motorsports' annual $20XX Challenge, where you can spend no more dollars on the car than the number of the year.

If you want to start with a worn-out car, make sure that any part of the car you plan on keeping is working reasonably well, and plan to have it off the road for long stretches of time. It is possible to drive a car while adding minor mods, but the classified ads that claim you can "drive while restoring" usually turn out to be empty promises. A car that does not run may be a great deal if you were planning to swap engines anyway, but a free car may not be a bargain if you spend more money trying to repair a rusted-out body and frame than what it would cost to start with a car in good shape. The most effective way to permanently fix rust damage is to cut out all the rusty metal and weld in new material, which is quite costly unless you can do this work yourself. Major body repairs are particularly expensive, and missing trim pieces can be extremely difficult to track down for older cars. As a general rule, cars with extensive body damage should only be bought to be stripped for parts and scrapped unless the car is rare enough to justify the expense of restoring it.

Given how many options you have when picking a performance car, choosing a car that meets your needs can be a little confusing. Going to local car shows may turn up cars from the 1930s parked alongside cars built just a few months ago. Attend a local race, and you may see front-wheel drive, all-wheel drive, and rear-wheel drive cars running head to head. Clearly, there are many eras and many drivetrain layouts that you can use to make a viable performance car, but they will all behave very differently.

How Old?

Cars from different time periods have different strengths and weaknesses. Different eras produced cars with different "personalities." You may have already decided that you want to have a car with the styling of the '60s muscle car or the latest high-tech

They still build them like they used to: Vic Edelbrock Sr.'s 1932 Ford is set up the way it used to race on salt flats and dry lakes. (*Photo courtesy Edelbrock.*)

performance. If you do not have a particular era in mind, here are some pros and cons of different car eras.

Cars built in the 1950s and earlier tend to get built with various degrees of updating. If you are buying a car for performance, you probably would not be interested in restoring a car from this era to keep it exactly the way it left the factory. Some hot-rodders like to build older cars with the sort of speed parts you could find in the 1950s. Tracking down the performance parts used in that era takes time and money, but you can still find some sources for parts for the more common engines in this era.

Street rods tend to do much more extensive updating. While some street rods use flathead Fords or other engines from the era when the car was built, it is more common to swap in a modern engine and transmission. Street rods often use modern suspensions and brakes, too. Many street rods are essentially high-powered modern cars wearing vintage sheet metal—or sometimes modern fiberglass bodies too. Some street rods have never even been real production cars, but were pieced together entirely from reproduction parts and modern running gear.

If you keep the original engine or suspension, you probably will not be able to find repair parts at the local parts store or junkyard. Instead, you will probably have to source parts from specialty shops. Parts generally will not be as cheap as ones for later cars, whether you are looking for performance or repairs. This is not the easiest way to get started in cars, although it is certainly not an impossible one.

Cars from the '60s can be cheaper and more practical, depending on where you live and what car you want. If you must have a '70 Chevelle SS, '65 Mustang Fastback, or other extremely desirable muscle car, expect to pay a fortune. You may not be able to get a good deal on '60s cars if you live in northern states where road salt has caused many of them to rust away. If you live in the South or California, though, there are still some cars from this era to be had for $2,000 or less in rough but restorable condition. Often, less popular cars will have as much performance potential as the more collectible

This '40 Ford street rod has more modern touches, from the wheels and tires to a Chevrolet V8 with modern speed parts. (*Photo courtesy Edelbrock.*)

'60s-era Chevelles are not cheap, but they are quite easy to work on and have a huge aftermarket.

choices. You can drop any engine that will fit into a '66 Mustang into a '66 Mercury Comet and use Mustang suspension bits on it too, but Comets do not cost nearly as much as Mustangs.

The muscle cars of the '60s have two great strengths: large engines and fewer emissions laws. In many states, cars from this time period are completely exempt from emissions testing, allowing you to put all sorts of things under the hood without worrying about whether it is legal. Many of these cars feature roomy engine compartments and are very easy to work on. Performance parts are extremely cheap and easy to find, at least for the more common engines from this era. This means that you can build a car from this era with 400 hp or so without spending a fortune, and do much of the work in your own garage. Furthermore, cars from this era are surprisingly light for their size, since they do not have as much sound deadener or luxury items as later cars. Nor do they have as much safety reinforcement to protect the driver in a crash.

You may be wondering just how reliable a 40- or 50-year-old car can be. My experience has been that you can use a properly maintained '60s car for your daily commute, but probably not your only means of transportation. It will need a couple of repairs and some basic maintenance yearly. The key is proper maintenance. Older cars can and will require more attention and minor adjustments. If you postpone fixing a problem or try to jury-rig a repair, look out. I once bought a Triumph Spitfire and had the dashboard catch on fire the first time I turned on the headlights. Somebody had improvised electrical repairs under the hood with lamp cord and no clue about how to correctly splice wires together. Fix older cars correctly, and they will reward you with dependable service.

Cars from this era are not without their drawbacks. They are not the most practical cars for everyday driving. Compared to modern cars, they are noisy and brutal. Handling and braking are not up to modern standards, although with some work and money you can correct this. Classic muscle cars also have a well-deserved reputation for their insatiable thirst for gasoline. One other issue is the insurance. While liability-only insurance is often cheap, getting comprehensive coverage on a daily driven classic that will actually provide enough money to fix it is very difficult. You may want to opt for a special collectors' policy that limits how often you can drive the car in exchange for a good price on comprehensive coverage. It is more practical to choose a '60s-era car as a second car, rather than as your main means of transportation.

In some ways, the '70s were an extension of the '60s. Unfortunately, the cars from this era were saddled with strict emissions rules that began to roll in around 1972. The smog equipment from this era often both hurt power and made the engines more prone to misbehavior. Today, cars from this era can often be a way to get a cheaper version of a '60s model. Most people looking to hop one of these up either look forward in time to install a later engine, or reach back to the past to build an engine as lacking in emissions controls as one from the '60s. Cars from the mid to late '70s are heavier and often less prestigious than older muscle cars, but if you want a cheap, rear-wheel drive car that is relatively easy to work on, you might want to consider one.

If you are just getting into American muscle cars from the 1970s or older, keep in mind that most collectors and enthusiasts prefer the two-door models to four-door sedans. Buyers at the time considered four doors less sporty, and nearly all of the people who ordered high-performance packages chose two-door hardtops, convertibles, or coupes. Consequently, two-door models today have more prestige and resale value. If you happen to get a good deal on a four door from that time period, however, it can perform just as well with the right parts. Recently, there has also been an interest in modifying classic station wagons. Four-door sedans may go up in price and mainstream

The cars of the '70s may not be as glamorous as the '60s, but they are much more affordable and just as easy to work on. (*Photo courtesy DIYAutoTune.com.*)

In the '80s, car manufacturers and hot-rodders found they had to do more with less. This Dodge Charger has a turbocharged, fuel-injected four cylinder driving the front wheels.

appeal as older sheet metal becomes rarer, but chances are they will still lag behind two-door models.

Many enthusiasts consider the modern era of performance to have started in the 1980s. Many of the trends started then continued on through the '90s and right up to the present. As manufacturers began experimenting with smaller engines and fuel injection, they also began to find new ways to build performance cars. Although turbochargers had appeared on cars as early as the 1962 Oldsmobile Jetfire, the new computer-controlled engines let manufacturers build far more predictable and reliable turbocharged engines. Aluminum engines and dual-overhead cam heads, which had also previously been specialty items, were mainstream by the '90s. Modern cars are often more comfortable and practical than cars from past eras. Many of them offer improvements in handling and braking, too. The biggest drawback when working on one is likely to be its complexity. Compared to earlier cars, modern cars have much more cramped engine compartments. The fact that engine controls are more electronic and less mechanical means that tuning requires somewhat different skills.

Drivetrain Layout

Today, you can find cars with front-wheel drive, rear-wheel drive, or even all-wheel drive. Although most cars have the engine at the front, a few cars mount the engine behind the driver. When the engine sits behind the rear axle, this is known as rear-engined. Cars that have the engine behind the driver but ahead of or on top of the rear axle are known as mid-engined.

The most common cars on the market today are front-engined, front-wheel drive. The usual reasons for this are low cost, light weight, and fuel economy. With very little distance from the wheels to the engine, a front-wheel-drive car has less friction getting power to the wheels than a front-engined, rear-wheel-drive car. With the engine over the driving wheels, a front-wheel-drive car gets good traction in rain or snow. Unfortunately, this traction advantage does not apply when you try to put a lot of power down on dry pavement. When a car accelerates, its weight shifts backward, causing the front wheels to have less traction.

Although front-wheel drive is not ideal for handling, there are many front-wheel-drive cars with excellent handling anyway. This usually comes from a combination of light weight and good suspension tuning. For a novice racer, front-wheel drive can be a bit more forgiving if you put power down too soon while coming out of a turn. Too much power can spin the front wheels, but at least it won't spin the whole car.

Many traditional performance cars are front-engined, rear-wheel drive. Not only does this make for better traction in a drag race, but many rear-wheel-drive cars have a better weight balance than front-wheel-drive cars. Some are still nose heavy, however, and a rear-wheel-drive car with most of its weight on the front can be difficult to drive in snow. If you give one too much power while cornering hard, you can cause the rear wheels to lose traction, making the tail end whip around. This may look spectacular in the hands of a skilled driver, and can definitely be fun, but it is not the fastest way around a corner. In most cases, the quickest way to toss a rear-wheel-drive car around the corner is to only hang the tail out a small amount and gently get on the accelerator once the car can handle the power.

The front-engined, rear-wheel-drive layout has another advantage besides its traction. This layout is often the easiest to work on, as the engine seldom spans the entire width of the engine compartment. Many of these cars have a good amount of room under the hood to allow for access or to accommodate wild engine swaps. There are a few exceptions with either a wide engine in a narrow car, such as the '70s Chevy Monza with V8 power, or an engine set back so far it is under the windshield, like the fourth-generation Camaro. For the most part, however, having the engine up front and the driving wheels in the back is the easiest design when it comes to tinkering.

Some popular performance cars have all-wheel drive. Although these cars were originally designed for rally racing on dirt roads, all-wheel drive can help on pavement, too. Sending power to all four wheels can help both starting from a standstill and with putting down power when accelerating out of tight turns. The only downside is weight, as all-wheel-drive cars tend to be heavier than two-wheel-drive versions of the same car.

Many trucks, and a few rare cars, come with a part-time four-wheel-drive system that connects the front and rear wheels with a clutch. This works well on mud or dirt, but should not be used on dry pavement. The four-wheel-drive systems used in performance cars like the Mitsubishi Lancer Evolution use a device called a center differential to allow the front and rear wheels to turn at different speeds, making the system suitable for use on paved roads. You'll learn more about center differentials in Chapter 8. Virtually any performance car with all-wheel drive will have a center differential.

Some people have claimed that vehicles described as four-wheel drive always lack a center differential, while any car marketed as all-wheel drive has one. This is not true; manufacturers have used these terms interchangeably. Some cars, like the rare Ford Tempo All Wheel Drive, have badges that say "All Wheel Drive" but do not use a center differential and must have the four-wheel drive activated by a switch. A safer rule is that almost no vehicles that need to be switched into four-wheel drive have center differentials. If you can turn the four-wheel drive off, the system is not going to be good for street performance.

Last, there are mid-engined and rear-engined cars. Except for some all-wheel-drive Porsches, these are usually rear-wheel drive. Such cars have most of the performance advantages of both rear-wheel drive and front-wheel drive. These often have what physicists and engineers call a low polar moment of inertia, which means that their weight is distributed in such a way that they can turn more rapidly. They are, however, somewhat tail-heavy. Some cars with this layout have an alarming tendency to swap ends if you drive them carelessly while trying to corner at the car's limits. This is especially a problem with rear-engined cars. If you are willing to deal with the challenge, however, the reward is often well worth the trouble. This category includes some of the best handling cars ever built, such as the Porsche 911 and the Toyota MR2.

Trim Level

Sometimes, you may find yourself trying to decide between different performance options or engines on the same model. Usually, the choice is between a cheaper base model and one built for more horsepower and handling. If maximum performance is your goal, the best choice is usually to get a head start by choosing the performance model. To make an Integra LS perform like an Integra Type R will typically cost you considerably more than the extra price paid for buying a Type R in the first place. While getting a car that already has the most factory performance options is usually the best choice, there are a few cases where this rule was made to be broken.

One of the best reasons to break this rule is if you are planning an engine swap to make an engine and car combination the factory never offered. If you dream of building a Civic powered by a K24 from an Acura RSX, it does not matter very much what engine was originally under the hood. Civics never came with this engine from the factory anyway. There are also a few cases where the factory made a large number of a particular engine, and a large number of a particular car, but only installed this engine in a tiny fraction of the cars they built. For example, only 80 Dodge Darts left the factory with a 440 Wedge under the hood, but this engine was fairly common in other Chrysler products. It is a lot cheaper to install a 440 in a Dart yourself or buy someone else's completed project car than to track down an original 440 Dart.

Some tuners like the challenge of building a car with a less popular engine. Setting your car apart from the pack can be a worthy goal, but keep in mind that most gearheads are not following the pack merely due to herd instinct. Trying to go fast in a Lancer without a turbo or a Camaro without a V8 will get you noticed, but it will also take considerably more time and money than the more popular choice. If this is your plan, it pays to go all out. A non-turbo Lancer with serious enough internal engine work to take on a Lancer EVO will get respect, but one with just a few minor bolt-on parts is not in the same category.

Another case where you might want to start off with something less than the highest performing model is if you are competing in organized racing. Sometimes, a series where you want to compete might not allow the top-of-the-line version of your favorite car. For example, if you want to go road racing, an '80s Mustang can only compete in the relatively affordable Improved Touring B class with the standard 2.3-liter four-cylinder engine. The Improved Touring B rules do not allow running a Mustang with

Turbocharging a six-cylinder Dodge Dart isn't nearly as easy as swapping a V8 into one. This is only the sort of thing you'd want to consider if you're looking for a challenge instead of a cheap way to go fast. (*Photo courtesy DIYAutoTune.com.*)

the 5.0 or the turbo version of the 2.3. Building a wallet-friendly car for sanctioned racing sometimes calls for avoiding the high-performance model.

Of course, maximum performance may not be your goal at all. You may be mostly concerned with gas mileage and insurance rates. In such cases, starting with the most economical version and just adding a few minor bolt-on parts makes perfect sense.

These goals are all somewhat different from building the fastest car you can with a minimum of hassle and a limited budget. If you simply want the fastest car for your money and do not plan any major work like swapping engines, starting with a car equipped with all the performance options is usually your best choice.

Budgeting Money and Time

Your choice of car and your choice of budget should fit together. It goes without saying that a $5,000 budget is not realistic if you want a Porsche 911, but going to the opposite extreme of a cheap car and a high budget can also come back to haunt you. Sink $80,000 into building the ultimate four-door Plymouth Volare, and few people are likely to understand, either when seeing it at car shows or looking to buy a hot rod. Building a high-dollar project using a cheap car is usually only a good idea if you absolutely must have that specific car and do not plan on selling it later, or if you've picked a model that a lot of other gearheads will also be willing to spend serious money on.

You can't always base your estimated budget simply on the cost of the parts you plan to install. This does work for simple parts that come in a kit with everything you need. More complicated installations, however, tend to require unexpected tools and small parts to make everything fit correctly. If you are planning any work that involves large-scale disassembly of your engine, swapping in any major junkyard parts, or custom building your mods, a common rule of thumb is to expect to spend one and a half to two and a half times the total cost of the parts you expect to use.

Time can be even more difficult to budget. Even professional builders who prepare the cars that appear on magazine covers sometimes have to deal with unexpected delays. Unexpected problems can make a part installation take two to four times as long as one might expect. The most frequent hang-up when working on cars at home with hand tools is stuck fasteners, but bad instructions and dropping tools or bolts into inaccessible spots under the hood can sometimes give stuck fasteners a run for their money. Suspension mods are often the most troublesome, since mud and salt often cause nuts and bolts to freeze solidly to each other.

This assumes that you are working on a simple "bolt-on" part, one where you can theoretically unbolt the original part and bolt the new one in place without disturbing many other parts. If you need to make trips to the store for extra parts or have to get the services of a machine shop to complete your project, these can add even more delays. Sometimes even installing a bolt-on part may require a trip to the machine shop if you break a bolt that is threaded into part of your engine. It is often a good precaution to have some other means of transportation when working on a car in case you need to get more supplies. That is one advantage to having a second car. If you have only one car, it's a good idea to invite a friend over to help with your wrenching sessions. This way, you and your friend have another vehicle ready if you need to go and get unexpected parts.

Setting Goals

Some goals are easier to set than others. When it comes to appearance, you may have a clear idea of how you imagine the car looking. When it comes to mechanical work, though, you may need to make some more detailed choices. These can be tough to make when you are not already familiar with what is out there. But there are some questions you can probably answer right away if your plan is to start with a particular car and make it into your ideal version of it.

When it comes to suspension, what is most important to you? Some owners just want the look of a low stance, while others want a car to corner as hard as possible. Lowering is not always the best way to make a car handle. Other priorities you might have include traction for a drag race or ride comfort.

Engine buildups tend to fall into four categories. One is simply bolting on a few parts to the stock engine. This often gives a little more kick without compromising reliability or requiring a lot of time. Some people use forced induction—a category of mods that includes turbochargers, superchargers, and nitrous oxide. These can more than double an engine's power, but are expensive and can make an engine less reliable unless you take the right precautions. Other engine builders prefer to add internal engine work without forced induction, for what is known as a naturally aspirated or "all motor" buildup. These engines typically make more power, but often require revving them harder to access the horsepower. One last option is to ditch the original engine altogether and replace it with a more powerful one. You will need to decide how much you can spend, how much work you are willing to do, and how much practicality you need for street driving.

You may have a different approach, with the goal first and the car later. If you want to compete in a particular type of racing, your best bet is frequently playing follow the leader: Start with a racing series and class where the competitors' budgets are in the same range as your own, and see what types of cars and what mods typically do well. It's usually the safest bet to put together a copy of one of the cars that does well. If you're feeling an urge to be different, you might want to find a car that you think could compete in the series and research its potential, but often this can be the harder way into the winner's circle.

The later chapters will cover how to pick the right parts for these goals.

Dealing with Dealerships, Laws, and Other Pesky Issues

Modifying a street car involves several legal issues. The first is whether installing the mods is actually street legal. This can involve several different federal, state, and local laws. The federal government sets minimum emissions standards, but enforcing emissions laws is normally in the hands of the states. The only federal laws a typical enthusiast is likely to deal with directly are the laws regulating lights and catalytic converters.

State emissions laws may be enforced in different ways. If you are not in a heavily populated area, they may not enforce emissions laws at all in your area. Some simply check to make sure that a catalytic converter is still there in the exhaust and go straight to testing the tailpipe emissions, a test that most well-tuned and well-maintained cars are likely to pass. Cars built in 1996 and later have sophisticated enough onboard diagnostics that an inspector may just connect to the control unit on the car, use that to scan if everything is in working order, and let the car through if it passes that test. Other states have a visual inspection, where the inspector checks to make sure that every visible part under the hood is either the original part installed from the factory or an approved replacement. Since California has the strictest inspection laws, a part approved for use in California is likely to be emissions legal anywhere in the United States. Parts approved by the State of California have a CARB EO (California Air Resources Board Executive Order) number to show that they are legal.

Emissions are not the only issue. Noise laws are common. Most states set a maximum sound level, while more draconian areas may have laws stating that the car may not be louder than when it left the factory. Some states may have other laws, such as specifying that cars may not be lowered beyond a certain point or forbidding use of nitrous oxide on the street. Check with your local laws when planning a project. Your local library will probably have a copy of the state laws in the reference section if you want a definitive answer.

The second problem you may encounter is with the dealership if your car is still under warranty. Some dealers give people the impression that they can completely revoke a warranty if they catch you putting aftermarket parts on your car. This is not entirely true; a dealer cannot refuse to fix a problem with your ignition switch just because your car has an aftermarket exhaust system on it. To deny warranty coverage, a dealer must show that a part you have installed could reasonably have caused the damage. If you bring in a car with an aftermarket turbo kit, it is quite reasonable for the dealer to insist that you pay for the blown head gasket yourself. Aftermarket parts may be covered under their own warranties, although these usually will not cover the labor required to replace the part.

Dealers have one loophole that they may use for denying warranty coverage even when parts you have installed have nothing to do with the damage. They may argue that you have abused the car and that the parts constitute proof that you drive the car hard. Some dealers have even been known to track down evidence that you have raced the car by searching online lists of race results. These claims may work for drivetrain parts, but even the arguments about abuse cannot be used to deny warranty claims if the part is not likely to break from rough driving.

Ending a Project

When planning what parts to install on your car, keep in mind what you plan to do with the car when you are done. If the car is leased and must be brought back in original condition, you will need to stick with easily removed mods. Adding custom paint, swapping engines, or welding in a roll bar can be very expensive to undo at the end of the lease. If you must lease, you would be better off adding only a few bolt-on modifications or modding a second car.

Selling a modified car is not always as troublesome as dealing with a lease, but mods often make selling a car more complicated. Do not expect to get back all of the money you spent on your mods. A buyer looking for daily transportation may consider the mods to suggest that you have abused the car. A fellow enthusiast may be more interested, especially if the car has mainstream appeal and was built well. If you are currently looking for a project yourself, getting a car with mods already installed can often be a cheaper way to get these mods than buying them yourself.

Finding a buyer who is willing to pay extra for a modified car depends on the popularity of your car and the quality of your mods. A classic Mustang with period-correct speed parts is likely to find a buyer. The same is true for cars prepared for popular racing classes like Spec Miata. More offbeat cars can take longer to find a buyer, and may need to be priced lower. No matter how much time you have put into tweaking a Geo Metro or how fast you have made it, the number of people who want a fast Metro will be smaller than the number who want a fast mainstream car. With muscle cars, keep in mind that four-door versions are often considerably less popular with collectors than their two-door counterparts. Of course, resale value may not matter very much if you plan on keeping the car indefinitely or want a particular dream car badly enough that you do not care about its resale value.

Sometimes, you may have to sell a project before you can finish it. You may find you have run out of time, money, or your landlord's approval, and find you must get rid of a half-running car immediately. Selling a project car that does not run right is likely to be a big hit to the wallet, as it is likely to go for only a fraction of what you paid for it. In this case, consider parting the car out, stripping the useful parts, and selling them one by one. This can definitely be a painful choice, but it is often likely to get back a larger amount of your original spending than simply selling the car outright.

Car Events

Even if you don't pick an event and build a car around it, sooner or later you'll want to show off what your project car can do. There are several different kinds of events and places to show off both your car's looks and its performance. Here are some of the more common events.

Drag Racing

The standing quarter mile is one of the most American of racing formats. Line two cars up against each other, give them the green light, and see who gets to the end first. Dragstrips can be found across the country and usually hold "grudge race" or "test and tune" nights most weeks where you can race any car that will pass a basic safety inspection. Admission can be as low as 20 dollars for a complete night of racing. A dragstrip with an active group of racers is likely to draw a wide variety of cars, including street cars from several decades, motorcycles, and sometimes even miniature dragsters with lawnmower engines.

You don't need to worry about whether your car will be embarrassingly slow, unless you have already been bragging about how fast your car is. People bring all sorts of cars to the dragstrip. Some racers bring their commuter cars or winter beaters on occasion just to see what they run. Others bring a car before they start any work on it to get a complete before and after picture of its performance. Some simply bring slow cars to compete in bracket racing.

Bracket racing is a type of drag racing that is often compared to *The Price Is Right*. Bracket racers attempt to predict how fast their car can take to reach the end of the dragstrip after making a few test runs. This prediction is known as the dial-in time, and is written on the windshield with shoe polish or an erasable marker meant especially for writing on glass. Instead of the starting lights counting down at the same time for

Run what you brung, and hope you brung enough! This Nova drag car certainly looks like it's brung enough for the loosely organized drag races at Atlanta Motor Speedway.

both cars, the slower car gets a head start equal to the difference between the predicted times. Whoever gets to the finish line first without running faster than the dial-in time or starting before the green light wins.

The set of starting lights is a complicated setup known as a Christmas tree. Each side has two small yellow lights at the top to indicate that the car is in position, called the pre-stage and stage lights. The countdown normally begins when both of these lamps are lit on each side. Below the small yellow lights are three amber lights, a green light, and a red light. The amber lights count down in half-second intervals. A half second after the lowest amber bulb lights, the green light turns on, and the car can leave any time after this light turns on—or, ideally, at the very moment the green light is illuminated. Leaving before the green light will light up the red light, showing that you have lost.

When going drag racing, be sure your car is safe. At a minimum, they will check to be sure your battery is strapped down securely, the wheels have all their lug nuts present and accounted for, and the radiator overflow runs into a catch can. The seat belts will need to be in working condition. They will require you to remove your hubcaps or wheel covers if you have any, and even if they do not require you to remove any junk from under the seats and in the trunk, this is a good idea. Faster cars will need extra safety equipment, such as helmets or roll bars. They will also require you to wear safe clothing. At a minimum you will need to be wearing cotton or wool pants and closed-top shoes.

Autocross

Autocross gives you a chance to show off what your car can do in the twisties, without a serious risk of damaging it. An autocross course is laid out using cones in a parking lot or on a go-kart track. Cars run one at a time around the course and their times are recorded. Autocross gives you a chance to learn car control and cornering techniques in an environment where there is nothing to hit but cones.

You may be wondering how fast cars typically go in an autocross. You will often see cars going fast enough to dive into a corner with one wheel lifted clear off the ground. The actual top speeds are not that high, seldom more than 65 mph, but roaring out of a tight turn at 65 mph in second gear with the engine screaming at redline and the tires howling in protest is a far more intense experience than cruising at 65 in fifth gear on the freeway. The small scale of an autocross course also means that turns come at the driver faster than in virtually any other sort of motorsport. Turns on a road course are often separated by hundreds or sometimes thousands of feet, while autocross frequently includes back-to-back turns spaced a mere 50 feet apart.

While drag racing has bracket racing to keep competition fair, autocross has car classes. Cars are divided into groups based on the car's performance potential and how many mods have been added. These classes keep a driver in a stock Honda from having to compete against a new BMW M3, or on some courses, keep the BMW drivers from worrying about the Honda CRXs. Many events will also have separate classes for novices and hold a driving school for beginners periodically.

Like drag racing, autocross requires safety inspections. Tech inspectors will check many of the same parts as the inspectors at a dragstrip, but will also check brakes and wheel bearings a bit more closely. An unmodified car in good, safe condition will probably pass. Roll bars and racing seat belts are not required for autocross, but all

Autocross puts a small race course in a parking lot.

drivers must wear helmets. Virtually all clubs that put on autocross events will have helmets available to borrow for those who do not have a suitable helmet. Also, drivers will need to help out with running the event when not driving. Usually, this involves picking up cones, passing out helmets and collecting returned ones, or that sort of thing. Check with your local club for more details about the rules and costs of entering, which are typically between $20 and $40 per driver.

Track Days

Those looking for more speed, but not quite ready to race wheel to wheel, can go to race tracks like Laguna Seca or Road Atlanta to participate in a variety of open track events. These typically run cars one at a time on a road course, often with an instructor in the passenger seat. Speeds are often higher than at an autocross, and there is considerably more danger of hitting something solid if you lose control. Drivers often are required to wear a racing fire suit, and depending on the event, cars may need to have an approved roll cage and other safety equipment installed.

Wheel-to-Wheel Racing

This is the sort of racing you often see on television, a pack of cars chasing each other around a road course or oval track at high speed. Wheel-to-wheel racing includes everything from NASCAR to the SPEED World Challenge to the events run at your

Track days let you run flat out on a road course, without much chance of hitting other cars. (*Photo courtesy Edelbrock.*)

Wheel to wheel at Virginia International Raceway. (*Photo courtesy DIYAutoTune.com.*)

local dirt oval track. Wheel-to-wheel racing usually calls for a car built especially to fit the rules of the class where you are competing. A tow vehicle is practically essential, as the chances of damaging a car in this kind of event are very high. If you cannot afford to wreck your car, you probably should not enter it in a wheel-to-wheel event.

Car Shows

Car shows come in all types and sizes. Some shows allow virtually any sort of restored or customized car. Others are restricted to cars from a certain brand or certain years. Some are small efforts held every month in a local parking lot. Others are enormous

Car shows offer a relaxed environment to show off your creation and see what other enthusiasts have brought.

national events like the Pebble Beach Concours D'Elegance, attracting thousands of cars. Shows may simply have cars on display, or they may have judged events. Cars may be judged on quality of modifications, how well preserved an older car is, or even on their stereo systems. Some shows include racing events, usually drag racing or autocross. There are countless opportunities to show off your handiwork, meet other enthusiasts, and see what others have done with their cars.

CHAPTER 2

Researching

Separating Marketing
Hype from Reality

Researching what you can do to a particular car can look like quite a daunting task if you're just getting started. Finding the right mods takes a general knowledge of how a car works, information about what is available, and the critical thinking skills to examine the claims made about these parts. Checking out what others have done to cars like yours can give you inspiration for where you want to take

Testing an engine in a dyno cell. (*Photo courtesy Edelbrock.*)

your project. A little research can also help you separate who is selling good mods from the companies that just want to make a quick buck from people who don't know any better.

There are two ways to know for sure if a mod works: testing it yourself or talking to people who have done the tests or seen the results. For anyone other than a millionaire playboy or a well-funded racing operation, testing every part made for your vehicle may not be practical. Even if you aren't doing very much firsthand testing, it pays to know how they test cars and how accurate the methods are.

Seat of the Pants

The seat-of-the-pants method, sometimes nicknamed the butt dyno, is definitely the most common way to test a car. While it is not the most precise, it definitely has its uses. Not many dealers would let you bring a car to a dragstrip on a test drive, after all.

The biggest drawback to this method is its lack of precision. While you may be able to feel if a car is accelerating faster, often a car may feel faster for reasons that have nothing to do with it actually being faster. More engine noise or just wanting to believe that your mods have made the car faster can also make a car feel as if it has more horsepower. It can also be easy to fall into the trap of believing, "It rides harsher, so it has to handle better." Seat-of-the-pants testing is often ineffective at measuring tiny changes best picked up with precision timing equipment.

Not everything requires precision measurements. Sometimes a subjective judgment is exactly what you need. This is especially true of handling quirks. "It leans too much

Sometimes, all the testing you need to do is just get out and drive the car and see how it feels. (*Photo courtesy Edelbrock.*)

around corners," or "The rear tires always feel like they're about to let go when I corner hard," are valuable lessons that you can learn if you push a car to the limits of its handling, preferably in a safe environment like an autocross. You may also notice engine issues, such as a slight hesitation between when you push the accelerator and the engine actually giving you more power.

Seat-of-the-pants testing definitely has its place. It can be useful to find your car's quirks, and sometimes to spot problems that should be solved before bringing a car in for more expensive performance tests. Just remember that precise measurements call for more precise measuring equipment—and that not everything calls for precise measurements.

Dynamometers

Horsepower can be measured in several ways. The most common way is to measure it directly using a device called a dynamometer, or dyno for short. Dynamometers fall into two major categories: engine dynos and chassis dynos. An engine dyno, as the name implies, is connected directly to an engine, with the engine not installed in the car. A chassis dyno, on the other hand, is connected to the drive wheels of a car, and measures the power actually put to the pavement. Since transmissions and other parts have friction that will reduce the power, chassis dyno measurements are lower than engine dyno measurements. The horsepower rating a car has from the factory is always measured on an engine dyno. The measurement from a chassis dyno is sometimes

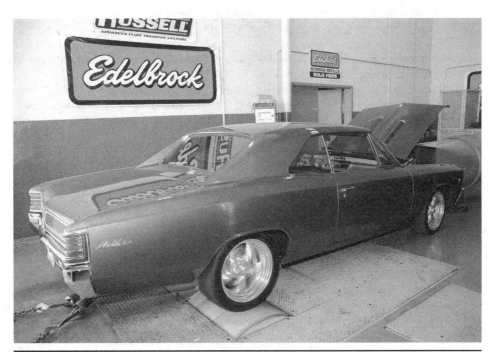

A chassis dyno measures how much horsepower you get at the wheels.
(*Photo courtesy Edelbrock.*)

called wheel horsepower, or whp. "Dyno tuning" refers to the process of making adjustments to an engine while measuring the results with the dynamometer. Dyno measurements are now available to average enthusiasts for under $100 for a set of three measurements, thanks to the development of inexpensive chassis dynos. Some companies even have portable dynamometers that they can bring to car shows.

HORSEPOWER, TORQUE, AND POWERBANDS

What exactly is this horsepower and torque that a dynamometer measures? This is not as simple a question as it may appear at first glance. A look at some debates on the Internet will reveal that there are a lot of people who don't understand the answers.

Horsepower is one unit engineers can use to measure power. To an engineer, power is a measure of the amount of energy a motor can supply in a given amount of time. Energy can be used for many things: moving a weight up a shaft, heating up a house, or making a car go faster. One horsepower is defined as enough power to lift a 550-lb weight upward by a distance of 1 foot in 1 second. The other frequently used measure of power is the watt, a metric unit. You're probably familiar with watts as a way to measure the power needed by electrical devices.

When it comes to cars, the faster a car is moving, the more energy the car has. The amount of energy is proportional to the weight of the car and the square of the speed. If air drag and friction stayed out of the picture, four times as much horsepower would let you reach a given speed in half the time. The power-to-weight ratio is the best way to estimate how quickly a car can accelerate.

The other measurement of an engine's performance is torque. Torque is best thought of as a twisting force, and is measured in a unit called pound-feet (lb-ft). If you have a 1-foot-long wrench handle and push on the end of it with 50 lb, you will apply a torque of 50 lb-ft to the nut at the other end. You could double the torque to 100 lb-ft by either doubling the length of the lever to 2 feet or doubling the force to 100 lb.

Both horsepower and torque are different at each revolutions per minute (RPM) point. A dyno operator will typically create a chart showing horsepower and torque as a function of RPM. These are known as a horsepower curve and a torque curve, respectively. The operator will also sum up the measurements by finding the peak levels of horsepower and torque, along with the RPM where the engine makes these.

For a given RPM, if you measure the torque in lb-ft, you can find the horsepower with this equation:

$$Horsepower = Torque \times RPM / 5252$$

One effect of this equation is that the horsepower and torque curves must always cross at 5252 if they are plotted on the same scale using English units.

Although the power-to-weight ratio is the best number to guess a car's ability to accelerate, the shape of the horsepower and torque curves can help you predict many things about how a car will behave. For example, a car that produces its maximum torque at 2,000 RPM will be able to take off hard if you suddenly floor it while cruising on the highway. One that produces its peak torque at 5,000 RPM will probably need to downshift several gears if you suddenly need power under the same circumstances. This car may also feel somewhat underpowered for normal driving around town, only showing signs of what it can do if you are willing to rev its engine without mercy. On the flip side, the engine that produces its torque down low is likely to run out of steam if you rev it up.

An engine cannot stay at its peak horsepower all the time without an extremely exotic transmission. A car that can sustain a high amount of horsepower above and below the RPM where it makes peak power will perform better than one with an equal amount of horsepower but where the power drops off sharply before and after the peak. This is related to the concept of powerband width, the distance in RPM between maximum torque and maximum horsepower. This is the RPM range where the engine will perform at its best. The wider the powerband, the better an engine can perform if your car is not quite in the right gear, and the easier the car is likely to be to drive. An engine with a narrow powerband will need many transmission gears and a lot of shifting to perform well, and is likely to lose a race to a slightly less powerful motor with a wider powerband.

During a dyno test using a brake dyno, or pull, the engine is run from a low RPM to its redline, with the throttle wide open and a brake being used to control the RPM. Smaller dynamometers used for lawnmower engines may use a brake similar to a car brake, but ones built for automotive engines typically use more exotic inductive, eddy current, or water brakes. Sensors on the brake measure the torque the engine produces at every 500 RPM, or sometimes in smaller increments. The dyno operator or a computer can then calculate the horsepower with these measurements.

Another sort of dyno is the inertial dynamometer. On an inertial dynamometer, the engine spins a large roller or flywheel, and a sensor measures how fast this spins. A computer uses these measurements to calculate horsepower directly, and then uses the engine RPM readings to calculate torque from the horsepower. Since an inertial dyno cannot hold the RPM steady, the typical way to use one is to conduct a pull by running the engine from idle to redline with the throttle wide open. Many cheaper chassis dynamometers are inertial, including the popular ones made by Dynojet, while engine dynamometers are nearly always of the brake type.

Dynos can be affected by several things other than the engine mods. Engines are a little sensitive to weather, so the altitude, humidity, and air temperature can change the measurements slightly. Different transmission gears have different levels of efficiency, too, so it can be possible to throw off results by as much as 5 hp if the car is tested in third gear before adding a part, and fourth gear after adding it. Often, a dyno may show a difference of 1 hp or so between pulls even if no changes are made, so performance changes this small are very difficult to tell from random errors. Individual cars may have differences of several horsepower due to wear, different build tolerances, and other factors, so ideally a comparison should be made between a car without a mod and the exact same car with the mod installed.

Another factor that can influence the horsepower reading is the dyno itself. Chassis dynos in particular tend to deliver wildly different readings depending on the manufacturer. For example, Dynapack chassis dynos often come up with numbers around 15 percent lower than ones made by Dynojet. If you're doing back-to-back comparisons, it's best to measure them on the exact same dyno to avoid differences in dyno readings overwhelming the differences the mods made.

When looking at the results of a dyno test, consider the entire combination of mods on the engine. Parts often work together as a combination. Replacing a muffler may not produce any gain if the rest of the exhaust system is original, but may result in a 10-hp gain if the rest of the system has already been upgraded with aftermarket parts and the original muffler was holding the system back. This is how it is possible to add three mods, each of which has been claimed to add 10 hp, only to find that the engine does not gain anywhere close to 30 more hp. Ideally, the engine should also have any tunable parts on it adjusted to match the newly installed mods. It is possible, particularly on older engines without computers, for a perfectly good performance part to reduce power if the engine is not correctly tuned to match the new changes.

The Dragstrip

A dragstrip is not just a racetrack; it's a measuring tool. Better yet, it's a very inexpensive measurement, and definitely one of the more enjoyable ways to measure your car's performance. The dragstrip has sensors in the pavement at regular intervals to measure your car's progress down the track. Four measurements are especially important: the reaction time, the 60-foot time, the trap speed, and the elapsed time (et). These four measurements show key information about the driver, the chassis, the engine, and the overall results.

The reaction time measures how quickly the driver was able to get the car moving. At most dragstrips, the timer for this will start when the final set of amber lights turns on and stop when the car begins to move. Since the green light turns on a half-second after the final amber light, a reaction time of 0.500 second is considered perfect, and anything less is a "red light"—meaning you lose. The car will usually not start to move immediately when you take your foot off the brake and hit the gas, so you will need to consider both your own reaction time and the car's reaction time when finding the right point to start. As a rule of thumb, the average driver can get an average street car going by starting as soon as the final amber light comes on.

The 60-foot time is a measure of how well your car can put its power to the ground from a standing start. Less traction will make for a longer time. Other things that can slow you down are trying to put too much power down too soon and spinning the tires, or starting it at too low an RPM and having to wait until the engine hits the point where it starts to make good power. Time lost here will further hurt your time down the track. Improving your suspension, tires, and launch technique—the way you work the brake, clutch, and throttle to get the car moving—will help you with this.

The trap speed is a measurement of how fast the car is going at the end of the quarter mile. This speed can be used to measure a car's power-to-weight ratio. The trap speed is not affected very much by how much traction the car has or how well the car gets off the line, only by how much power the engine has and how much the car weighs.

Handling Measurements

Handling is a bit harder to turn into a nice, clear number. There are some tests that are quite straightforward and repeatable, but no number can tell the whole story. Braking itself is fairly easy to measure, and is usually done by putting the brakes on as hard as possible without skidding from 60 mph and seeing how long the car travels before it comes to a dead stop. To be sure it can still stop when the brakes get hot, it is important to test the brakes several times, usually at least ten times for a street car. Cars that hit high speeds on a road course may need even more torture testing of their brakes to make sure they're able to hold up to track use.

One way to measure handling is to see how fast a car can complete a race course.
(*Photo courtesy DIYAutoTune.com.*)

Cornering can be measured with a slalom or a skidpad. On a slalom, the car turns rapidly between a line of cones at a regular interval, and someone with a stopwatch or optical timing equipment measures the highest speed the car can reach on this test without running over the cones. On a skidpad, the car drives in a circle, and the speed it can reach is used to calculate what is known as lateral acceleration. Lateral acceleration is measured in g's, and if you are in a car cornering at 1.0 g, you will feel as if you are being pulled sideways by a force equal to your own weight. Both slalom and skidpad tests are fairly standard, so a test in one magazine can be compared to another and is likely to be in the ballpark. (They can be affected by things like what sort of pavement was used and air temperature, so don't split hairs if the numbers are close.) Accelerometers such as the G-Tech can measure lateral acceleration directly, as well as braking and forward acceleration.

Sometimes, you will see cars tested on an autocross course or road course. These numbers make the most reliable comparisons if run on the same day, and since autocross courses are set up with cones in a parking lot, each test will give numbers that usually can only be compared to other numbers from the same testing session. Driver skill is also a key factor. Road course times are a bit easier to compare, but it's not very often that a magazine or other tester will be able to rent a well-known track like Laguna Seca or the Nürburgring for testing. Track tests can provide the best indicator of whether a mod will help in a real race.

Many measurements of handling, unfortunately, do not come up with useful numbers to compare. There is no easy way to measure how roughly a car will ride. There are no numbers to tell you if a car will give you an ugly surprise if you push it too hard. For such issues as comfort, responsiveness, and predictability, you will have to rely on the opinions of the people who have tested the mods firsthand.

Smartphones and Testing

The latest crop of cell phones combines a built-in global positioning system (GPS) and accelerometer with the ability to download and run complex programs. This makes them handy for measuring a car's performance, and there are several programs that can use the GPS and accelerometer data to calculate skidpad numbers and similar times to a dragstrip. The more accurate versions have the capability to tie into the car's electronics to record engine RPM and other data as well. These may not be quite as accurate as real dragstrip timing equipment or professional data acquisition, but the results can still make a useful comparison.

Researching Mods

Personally testing every mod for your car is not a very realistic option. Finding others who have tested your car and learning their results can give you an idea of what works without spending nearly as much time or money. It is up to you, however, to discern if your sources of advice are accurate and trustworthy. There are several good places to look.

Racers often like to hang out and compare notes with each other at amateur racing events.

Visiting Race Events

Racers have a saying: "The BS stops when the green flag drops." Attending amateur racing events gives you an opportunity to watch the green flag drop (or the green light come on), and watch the timing equipment separate the serious racers from the ones whose claims about their cars were nothing but BS. Visiting your local dragstrip on test and tune night may only cost a few dollars, while watching an autocross event is often free. In either case, you will be able to see the times these cars have run on display for all to see, and many drivers will be quite happy to show off what mods they have installed to get their times. Racers may get tight-lipped if real money is at stake, but at events run for not much more than bragging rights, well, many of the guys in the lead love to brag about what parts (and techniques) got them there.

Car Clubs

Virtually every car you can find, no matter how obscure, probably has an organization of owners somewhere. Some clubs hold regular car shows and meetings, while other clubs only exist in cyberspace. Both real-world and online clubs offer excellent opportunities to meet other people interested in the same car you have, giving you the chance to swap stories, learn what installing various mods involves, and compare what parts work.

Online organizations are fairly easy to find on the Web with a good search engine. Some use message boards, which allow you to write comments on the Web, while others use mailing lists, which allow you to send email to all the other members and receive email from them. Online organizations also will frequently keep articles submitted by members, covering everything from routine maintenance to aftermarket parts installations to creative things that can be done with junkyard parts. Many online resources will have a FAQ—a list of frequently asked questions. It is best to read the FAQ as soon as you join and find if any of your questions are answered there. Sometimes a few members may get upset about being asked questions already answered there, which is a common quirk of Internet communities.

Real-world clubs can be found in several different ways. Many will have a presence on the Internet and can be found with a search engine, but others are not online at all. Local newspapers will often carry information on local car clubs in their Events section. Car clubs sometimes also advertise at major car shows, and asking around at car events is also likely to turn up a few car club members.

Books

Think of this as mods for your mind. A comprehensive book like this one can provide an overview, but to fully cover all there is about every component and examine every car in the motorsports scene today would require a work the size of the *Encyclopedia Britannica*. There are many excellent books that provide detailed, in-depth looks at many areas of performance cars. Some cover specific parts of a car in depth, such as the engine, suspension, custom paint, or the details of designing a turbo setup. Others are dedicated to modifying particular cars, such as Mustangs or Civics.

Model-specific books are very useful if you can find one that covers your car. They often will offer extensive information about mods they have tested and proven parts

The book you're reading now can't contain everything there is to know about racing. There's a lot more out there for further reading.

combinations. Better yet, many of these books also cover low-buck tricks that you can do using junkyard parts, creative work with a drill and hacksaw, and other free or cheap mods. The best examples also include theoretical details.

You will definitely want a repair manual for the car you are working on. Haynes and Chilton both publish inexpensive manuals with many photo illustrations. If you are planning on anything beyond simple bolt-ons and basic maintenance, you'll also want a copy of the factory service manual. This gives you the same information that the dealer's mechanics have. These manuals cover far more than the aftermarket manuals, but often tell you to use special tools that you aren't going to find at the local parts store. Sometimes you may even get a blank look if you ask the dealership service department where you can buy a rare tool for a long-discontinued car. Often, the aftermarket manuals will cover how to accomplish these sorts of things with improvised tools—if they cover the repair. A Chilton's manual is usually thinner than a factory manual, and with more photos—not everything can make the cut.

Magazines

Magazines provide coverage of everything from race results, to feature cars, to individual mods. The more in-depth articles about individual mods may include an

One sign you're visiting a gearhead house is that there may be entire shelves dedicated to car magazines.

illustrated explanation of how to install them and a description of how the writers tested the parts. The test results are often quite informative, particularly if the magazine has been testing several competing parts against each other. Often the parts tested will have been donated by aftermarket manufacturers and advertisers in the hopes of getting a magazine to showcase their work. While not too many magazines are willing to bite the hand that feeds them, they're more likely to just not give bad parts any ink. It's almost unheard of for reputable magazines to create fake test results showing a worthless part adding horsepower.

Feature car articles can provide great examples of what can be done with a car, and often give quarter-mile times and horsepower measurements as well as a list of work done and pictures. Many of the feature cars, except in the most budget-oriented magazines, would be difficult to build cheaply. After all, glamor costs money. But it's often possible to learn about what mods work and pick up styling ideas from these feature cars.

Magazines often have their own websites, too. These sites may offer free copies of some of their articles. Most magazine websites also offer message boards where you can ask questions and get answers from the magazine subscribers, and sometimes even have the staff personally answer your questions.

The Internet

The Internet offers several other useful resources besides online car clubs and magazines. There is enough good quality information to teach you how to build race car electronics or carbon fiber speed parts on your kitchen table. Unfortunately, there is also so much bad information online that the Internet is sometimes dubbed the Misinformation Superhighway.

There are many reasons why misinformation springs up online. Some is simply put up by beginners who don't know any better. Other bad advice comes from scammers looking to peddle questionable products. You may even find characters who get their kicks pretending to own a much cooler car online than they can afford in real life, with equally made up stories about how they modded their cars. In a healthy Internet community, these imposters usually wind up sticking their virtual necks out too far after a while and get decapitated by a genuine expert.

Typing the name of your car into a search engine will often bring up countless personal home pages from people who also own a car like yours. Many of these pages will contain lists of modifications and stories about how the owners have tested out their cars. With no editors on the Internet, the quality of information there varies greatly. Luckily, most home pages tend to give you enough information to get a feel for the writer's level of expertise.

Online shopping websites can also provide a feel for what parts are out there for a particular car. Obviously, they are trying to sell you their products, but examining these sites can give you valuable information about how much aftermarket support is available for your car or engine.

Many racing organizations will also have a presence on the Internet. Racing message boards can be as useful as model-specific message boards in some cases. You will also be able to find the results of many racing events posted on the Web. Looking through the results for a racing class that uses lightly modified cars can be almost as useful as attending races in person.

Hired Gurus

Perhaps you are more interested in driving the car than in picking the right combination of mods yourself. If you want a heavily modified car and find selecting the right parts intimidating, there are several ways to have someone else do the design work and install a proven combination of effective parts. In some cases, you can simply take the car to a tuner's shop, let them know what kind of horsepower and handling you want, and have them do all the work.

Another option is to get a pre-packaged set of parts. One of the most common examples is to buy a crate motor, a complete engine with all the performance parts already on it. Then you or your mechanic can simply put it under the hood. In other cases, a tuner or manufacturer may have a complete package of engine or suspension parts, all tuned to work with each other.

Although having a tuner do the work can save you the trouble of researching the individual parts, you will have to do a bit of checking on the tuner's reputation. Magazines may provide insights into a tuner's reputation, but there are a couple media darlings out there who somehow manage to get good press while leaving a trail of broken parts and angry customers. Checking on Internet forums and local car clubs can

If you're not sure you want to pick out every last part for your engine, there's no shortage of turn-key engines out there with the parts chosen by experts. (*Photo courtesy Edelbrock.*)

often turn up these sorts of horror stories about the worse tuners and help you get good information about tuners too obscure to get very much magazine coverage. Race results are also a good way to spot a quality tuner—you can't fake winning.

Spotting a Bogus Mod

Mods come with many different levels of quality, but some are outright rip-offs. The first way to avoid such bogus mods is to have a good understanding of how your car works. For example, knowing that removing restrictions from your intake usually frees up more power should make you suspicious of any device that claims to make more power by placing a restriction in your intake. Some of the bogus mods will make claims that go even further and completely defy the laws of physics.

Another tip-off is that bogus mods frequently cost absurdly less than real ones that are described with the same terms. For example, if a performance chip typically costs $200 for a certain car, and you see someone advertising a "chip mod" for only $20, something's fishy. On a few cars where anyone can crank out performance chips in a

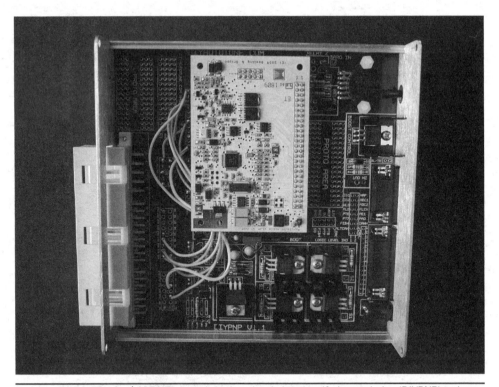

A plug and play ECU for $425? The catch is that the do-it-yourself plug and play (DIYPNP) arrives as a solder-it-yourself kit, and there aren't tunes out of the box for everything it fits. This could be a good deal if you're willing to put the work into it, but a ready-to-go plug and play system would probably be cheaper if you're paying someone else. (*Photo courtesy DIYAutoTune.com.*)

spare bedroom, it may simply be a copy of a professional tune or some amateur's effort at tuning a car himself, but on other cars, it probably isn't a real chip. Likewise, a device that will allegedly "supercharge" your engine for $70 is not likely to be a real supercharger, given that supercharger kits typically retail for $2,500 or more.

Inexpensive parts are more likely to be outright rip-offs than higher-priced parts, probably because most people willing to spend hundreds on a part are savvy enough to spot bogus parts. There are a few occasions where you can find a real mod for a significantly lower price, but in these cases there is usually a catch, which a reputable seller will tell you about up front. It may be a cheap Chinese copy of a part developed elsewhere. Or it may come with a considerable amount of assembly required compared to other products.

Real performance parts often require extensive testing to make them work well. Companies that have done this will proudly share their results, with power measurements and often a graph of power versus RPM compared to the original part or one made by a competitor. Pay close attention, though. If the horsepower gain they advertise is at a particular RPM point instead of saying they are comparing peak horsepower, you probably won't see that much of a change in peak horsepower numbers. Also, note what other parts were on their test engine. A part that adds 2 hp to an 800-hp race engine probably will not add 2 hp on a 150-hp street engine. Pages of fine print are not necessarily signs that a part is a rip-off, just that you need to know what combination gave the engine this power. A part that gives a slow-turning V8 "30 more horsepower at 6,000 RPM" may only give you an additional 5 hp at the peak, but if you want to help your motor keep pulling hard all the way to redline, it might be just what you need.

On the other hand, a promise of "0 to 20 hp" suggests that the manufacturer has not done much if any testing, and you can probably guess which end of that range it will give your car. The lack of real testing also makes it much easier for companies selling bogus mods to offer their parts for a huge variety of cars. If all you have to do is figure out how to mount your part on the car and don't have to worry about all those pesky tests to make sure it provides a boost of horsepower, you can sell your part for almost any car. Very few serious performance parts are truly universal, although good motor oil and mufflers can come close.

Sizing Up a Supplier

Mods based on sound theories can still vary widely in the amount of testing done and the quality with which they are built. There are several ways to find out more about whether a brand you are considering is a top-notch product, simply good enough, or a brand to avoid.

First, take a look at the product line the company offers and whether it is involved in any sort of racing. If a company primarily makes appearance mods with a few basic, low-priced mechanical parts, and its products almost never turn up on a race car, you can be certain that their performance parts are also likely to be aimed at an audience who is more interested in how the parts look than how they perform. The flip side is that a company that only builds parts for serious racers may only have parts on hand that are too hardcore to use on the street. Some companies make both racing and street parts. You will have to decide if the manufacturer's expertise matches your goals for your car.

Ask around, both at clubs and on the Internet, to see if you can get any personal opinions of the mods you're considering. Be honest about how you plan to drive the car, particularly when it comes to suspension mods. A top-notch racing suspension can easily be way too punishing for the street, and a suspension that feels acceptable to a street driver may turn out to have unpleasant quirks if pushed to the limits on an autocross or road course. Ideally you will want comments from people who have used the same mods on the same car you have for the same purpose you plan, but comments about other products from the same company can provide a useful picture of the manufacturer's overall standards.

Magazines will often test parts. These can provide both a detailed idea of what it takes to install the parts and measurements of the mod's effectiveness. If a magazine routinely uses a certain company's products on their project cars, this likely means that they have found the products to work well in the past, or that they have a good relationship with the manufacturer (which might mean the manufacturer buys a lot of advertising and is always happy to send them sample mods for their project cars), or both.

Last, some parts are simple enough that you can easily examine them in a store and tell if they are likely to work well. These parts, and what to look for, will be explained in later chapters. Two of the more common examples of easy-to-check mods are strut tower bars and straight-through style mufflers.

Doing Your Homework to Ace the Tests

It can definitely pay to do your homework before installing a mod. A sound knowledge of how a mod should work and some knowledge of the manufacturer's reputation are a great place to start. If you can get accounts from people who have installed a part just like the one you want, this is ideal. Install several mods and test them out at the races or dyno shop, and it won't be long before beginners start asking you for advice.

CHAPTER 3

Wrenching

Tools You'll Need and How to Use Them

With so many tools on the shelf at the local auto parts store, you may wonder how many of them you will need to work on your car. The good news is that many mods can be installed with a fairly small set of basic car tools. Other, more specialized tools are something that can be added to your collection when you need them. Better yet, many parts stores will lend you specialty tools with a deposit, so you may not need to own tools that you seldom use.

Jacks and Jackstands

Working under a car is not safe if the car is held up by cinder blocks. If you need to lift the car to work under it, make sure you are supporting it safely. You will want a jack, jackstands, and wheel chocks to hold the car in place. Lift the car on a solid concrete or asphalt surface, and never crawl under a car that is held up by a jack alone. I've lost count of how many times I've seen jacks break, topple over, or just gradually lower the car back down when they shouldn't. Instead, use the jack to lift the car, and support it with the jackstands. The owner's manual will show where to place the jack to lift the car.

Jacks come in several varieties. The most common types are the floor jack, scissor jack, and bottle jack. There are a few other designs, most of which tend to come with cars as original equipment but usually aren't a mechanic's first choice.

Most cars today come with a scissor jack for changing tires on the side of the road. This jack will lift the car, but is designed more for light weight and to fit in a small space than for ease of use. A light-duty floor jack can be picked up for around $20 and might be somewhat easier to use than a scissor jack, but these are somewhat fragile and not as easy to use as a full-sized floor jack. This price will also buy a heavy-duty version of the scissor jack that may be reasonably sturdy but can require a bit of effort. Full-sized floor jacks can run anywhere from $60 to several hundred dollars, but they make lifting a car considerably easier and are definitely something to consider buying if you intend to do a lot of work on your car. The bottle jack is a special-purpose jack. It's too tall to fit

You don't need a huge shop full of tools when you're just getting started. This toolbox holds most of the tools you'll need for your first car projects.

A sturdy, lightweight aluminum racing jack makes lifting your car much easier than a cheaper type.

Use a set of jackstands to support the car once you've got it up in the air. Never work under a car supported by just a jack.

under most cars, but can work with some trucks. They also come in handy for a few specialized jobs that don't actually involve lifting the car.

Wrenches

There's a reason gearheads often use "wrenching" as a synonym for working on cars. The vast majority of the work done in installing parts is removing and reinstalling the screws, nuts, and bolts that hold the parts of the car together. This is usually quite straightforward. Other than the number of bolts you must remove, there is little difference between the techniques you need to put on a cold air intake and the techniques necessary to swap cylinder heads. It just takes more time and effort.

There are several styles that have a place in the mechanic's toolbox. The most familiar style of wrench is a solid piece of metal with open-ended tips, each tip grasping a single size of bolt. These are useful for when you have to slip the wrench over a nut from the side, but can slip off, sometimes so violently that open-ended wrenches are sometimes nicknamed "knuckle-busters." Box-ended wrenches have circular ends with pointed holes, which make the wrench very unlikely to slip off, but can make it harder to reach a nut in some cases. Combination wrenches have one open end and one box end, making them the best choice if you are only buying one set of wrenches. Flare wrenches are a hybrid of the open end and box end, with a bit of the box end cut out. You use these on tube fittings like brake lines and fuel lines, where the extra grip of the

Combination wrench (top) and flare wrench (bottom).

box end can help avoid damaging the nut but you need an open end to slip over the tube. Adjustable wrenches are a version of the open-ended wrench that can fit several sizes, but can slip off more easily and are more likely to damage a stuck bolt.

Socket wrenches use a cylinder with a cutout in it (called a *socket*) to fit onto a nut or bolt and a separate handle that can be used with different sockets. Normally, these are used with a ratchet handle, which can make turning bolts considerably easier. The ratchet mechanism allows the socket to only turn one way, letting you tighten or remove

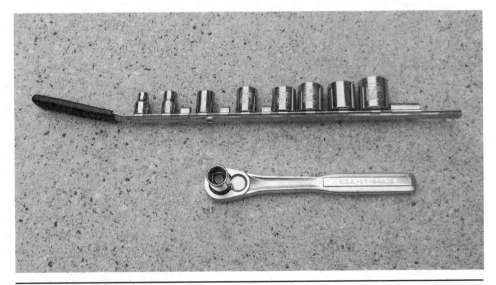

Sockets and a ratchet wrench.

nuts simply by swinging the handle back and forth while leaving the socket on the nut. Socket wrenches are definitely one product where you get what you pay for. A set of American-made sockets with a lifetime guarantee (or ones from Japan, Germany, or other countries not generally associated with cost-cutting nowadays) are considerably less likely to break or damage a fastener than a cheap set of Chinese sockets.

There are a wide variety of accessories that you can use with your socket wrench set. Deep sockets have more room for long bolts. Extension bars and U-joints let you reach otherwise inaccessible fasteners. A breaker bar is a long handle with no ratchet, and is just the thing for removing a stubborn nut. One other tool to use with a socket wrench set is a torque wrench. There are several sorts of torque wrenches, but all are designed to measure the tightness of a bolt. A torque wrench is essential if you need to change a cylinder head, replace spark plugs in an aluminum engine, or deal with bolts anywhere else that absolutely must be tightened correctly. Although their length makes it tempting to use a torque wrench to deal with a stuck fastener, this can damage the torque wrench. Use a breaker bar for this instead.

Professionals often use a type of wrench called an impact wrench. Instead of applying a steady torque to a socket, an impact wrench hammers on the bolt with sudden bursts of torque. This often works great for unfastening stubborn nuts and bolts, and you often don't need to secure the other end of the bolt as tightly. The trouble is the expense. Up until recently, impact wrenches required compressed air, with the expenses of air lines, compressors, and all. Recently, electric impact wrenches have appeared on the market, although good ones are pretty expensive and the cheap ones often don't improve much over a breaker bar. If you can afford one, a good impact wrench can make disassembly a lot easier. Note that impact wrenches need their own sockets; the sockets designed for hand wrenches are not made to stand up to the sort of pounding an impact wrench will dish out.

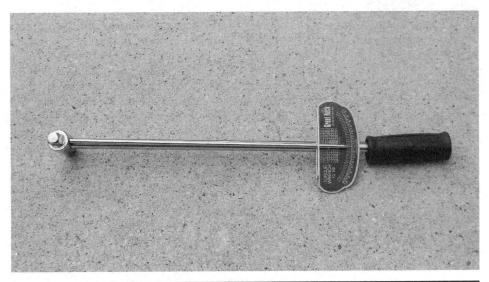

A beam-type torque wrench is not as easy to use as a click type, but it's tough, cheap, and works for most jobs.

One more sort of wrench is a lug wrench. You can remove the lug nuts on your wheels using the wrench that came with your car, or a deep socket and a breaker bar. However, a four-way lug wrench can be spun much faster, speeding up the time needed to remove or reinstall your wheels.

Wrenches come in two common standards: metric and SAE (inch). Virtually all Japanese cars, German cars, and most newer American cars require metric wrenches. American cars of the '60s will require inch tools. American manufacturers began to phase in metric parts in the '70s and '80s by designing new parts using metric units but keeping inch fasteners on parts until they designed metric replacements. Consequently, some American cars will require both sorts of wrenches. There are a few other very rare nut and bolt sizes out there. The only other one that normally shows up on cars is the Whitworth, used on some older British cars.

Screwdrivers

Most of the time, you will only need basic slotted and Phillips screwdrivers. Sometimes you will find parts that use other styles of screw heads, including hexagonal (Allen) and six-pointed stars (Torx). Sometimes you'll get the impression designers picked these oddball screws to make your life frustrating. If you're trying to work on something that was put on to satisfy some legal requirement, this might literally be the case, but usually there's another reason. These specialty screws can be tightened more and are less likely to be damaged by the screwdriver. Always use the correctly sized screwdriver for the screw to avoid damaging your fasteners. This is not always obvious, as you can often get a wrong-sized screwdriver to turn the screw, but the right size will let you apply more torque safely. Abusing screwdrivers by prying open paint cans or the like will also shorten their lifespan.

Other Basic Tools

There are a few other basics that everyone who wants to work on cars should have. One is a suitable set of gloves. Ordinary yard work gloves are not the best choice here. Gloves designed for working on cars are tight-fitting, but offer enough protection to avoid getting any nasty cuts if your wrench suddenly slips off while trying to turn a bolt. Most of these gloves are washable, too. Another product you will want for your hands is a jar of degreaser. Ordinary soap often does a poor job of removing the sort of grease you are likely to get on your hands while working on cars, even if you are wearing gloves.

Speaking of grease, there are several tools that are useful for oil changes and similar maintenance. A funnel is useful for pouring oil or transmission fluid, and you will need a drain pan to catch fluids. Greasing the suspension typically requires a grease gun. There are special tools designed to make removing your oil filter easier; my favorite is the chain wrench.

Pliers are a useful thing to have in your toolbox, particularly when dealing with clips and other small parts that must be grabbed firmly or squeezed. In addition to a standard set of pliers, you will probably want a set of needle-nose pliers for working in tight spaces and a set of locking pliers. Locking pliers such as Vice-Grips are one of the few tools that will grab a nut if it becomes too stripped to use a normal wrench. Since locking pliers will often scratch anything they lock on to, they are best used on parts that are already damaged beyond repair and need to be removed.

This chain wrench can remove almost any size of spin-on oil filter.

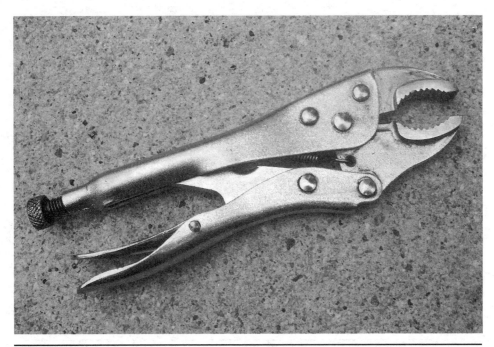

Locking pliers are great for grabbing stubborn parts.

Containers and Organization

Misplacing bolts and small parts can cause you no end of frustration. Airplane mechanics have a rule that whenever they remove a fastener, they place it in a small bag and hang the bag right next to where they removed the fasteners. While you may not need to be this organized, you will find it very useful to keep all the parts you have removed in a spot where they are not likely to roll away or get lost. Sandwich bags or magnetic trays can be very helpful here. There are few things more frustrating than trying to put your engine back together and not knowing where all the bolts from your project went.

Actually, there is one more frustrating thing when it comes to misplaced bolts. Like finding half a worm in your apple, finding that a small part has fallen into your engine can be a rude surprise when you try to start the motor. I've had to replace a distributor once because I forgot that I dropped a screw into it and attempted to start the engine. If you drop a small part, stop all work until you have recovered it—or at least made sure it's landed someplace harmless.

There are a few tools to help pick up dropped parts. Some have a magnet on the end of a rod. Others use a thin, flexible tube with a claw on the end. Having one of each can be very useful if you drop a socket or bolt into an inaccessible corner of the engine compartment.

This claw tool is great if you need to recover a bolt that's fallen into an inaccessible spot.

When Bolts Stick

The most common nuisance you will find when working on cars is that nuts, bolts, and screws sometimes stick. If you encounter a stuck nut, the first thing to try is penetrating oil like Liquid Wrench or Kroil. Pour some on the threads that are sticking, and give it an hour or so to soak in. If you are working on your suspension or someplace else where you expect to encounter stuck fasteners, you may want to apply penetrating oil to all the bolts the day before you start working.

If the oil does not work, you may need to use a torch. A propane torch may work, but an oxyacetylene or oxygen/MAPP gas torch will be more powerful. Heat the fastener until it just begins to glow red hot, and then let it cool. This can burn off contaminants and crack the rust that causes nuts to stick. Once it is cool enough to touch, try removing it again. Usually the fastener will have loosened up. It goes without saying that a torch is not the safest tool in the world, so if you are using a torch, be very careful where you point it and have a fire extinguisher handy. Some parts of a car are obviously ones you would want to keep away from a torch flame, like fuel lines, but other hazards are less obvious. For example, shock absorbers are filled with pressurized oil that expands when heated, and since modern shock absorbers are sealed, too much heat will take that word from "expands" to "explodes."

While good-quality tools can make it less likely that you will damage fasteners, sometimes this happens. One common sort of damage is when the corners of the head get rounded off. When this happens, it is extremely difficult to apply enough torque to remove the fastener. Your best bet in this situation may be to grab it with an appropriately sized Vice-Grip. If this still slips no matter how hard you tighten it, try heating it with a torch, as described earlier. In the case of Phillips-head screws, you can also use a file or grinder to cut a slot in them and use a slot-type screwdriver.

The most annoying stuck fastener is a machine screw that breaks off inside a part. Sometimes you can remove it by taking a left-handed drill bit and drilling down the center of the screw with a hand drill running in reverse. More often, things that cause a bolt to be so stuck that it snaps in half do not yield to common home tools. If possible, I usually deal with this by completely unbolting the part and taking it to a machine shop so I can pay someone else to deal with this stuck fastener. Usually the only reliable home fix is to drill out the bolt hole and rethread it using a Helicoil or similar insert.

Occasionally it will seem as if the parts are stuck together, rather than the fasteners. Nine times out of ten, when this has happened to me, I've found another little bolt somewhere, hiding out where another part kept it out of my view. There are a few other causes. Some parts are held in by a clamping force, and they may need a little prying with a screwdriver to open up whatever is doing the clamping. Few parts truly stick to each other; usually some other part is holding them back. There are a few exceptions, usually items like brake drums. In those cases, tapping the part with a mallet (preferably one with soft faces like rubber, plastic, or lead) will usually get the part to come off.

Presses

Some parts, such as wheel bearings, use a press fit. This means that the part is made slightly larger than the hole and needs a large amount of force to get it in or out. Although you can buy a press that will fit on your home workbench, most of the time it is easier and cheaper just to take the assembly with the pressed-in part to a machine shop.

Electrical Basics

Car electronics are not that hard to work with, but if you have not done much wiring work, here are some basics of electricity and car electrical systems. This is a very basic introduction, suitable for simple applications like getting power to new mods.

Wires transmit power using a flow of electrons. The electrons flow from the negative side of a voltage source like a battery or alternator to the positive side. The voltage source then moves them to the negative side so they can travel around the circuit again. Logic would suggest it should be the other way around, with electrons running from positive to negative. Unfortunately for science students, by the time scientists had figured out that electrons had negative charges, the terms were already too well established to change.

The rate at which electrons flow through a wire is known as current. Current is measured in amperes, or amps for short. If you must know, one amp means that 6,250,000 trillion electrons pass through a wire every second. When wiring up new parts, it is much more important to know that your wiring can safely deliver enough amps.

Amps are not the only unit used to measure electricity. One of the other important units is the volt, which measures how much energy each electron has. If you connect a device that needs 12 volts to 100 volts, it will get too much energy, and probably overheat and burn out. Connecting it to a 1.5 volt battery, on the other hand, will usually not supply it enough energy to work correctly. If you want to find the amount of electrical power a device uses, which is measured in watts, multiply the volts it needs by the amps it requires.

Most cars use what is known as a 12 volt, negative-ground wiring system. That means that the electronics are designed to be powered by 12 volts (more or less—it's usually somewhere from 13 to 14.6 volts when the engine is running) and that the negative terminal of the battery is connected to the body of the car and the engine block. Electrons in a negative-ground system run from the negative terminal of the battery, through the metal body of the car, to the electronics, and return back to the positive terminal through wires.

OTHER WAYS TO WIRE A CAR

A few older cars may use oddball systems such as positive ground or 6 volts. You probably do not need to worry about this unless you have a car from before the 1960s or some European oddballs. Parts meant for 12 volt systems will not work on 6 volts. Electrical parts meant for a negative-ground system can work in a positive-ground car if you take special precautions when you install them. If the part has a plastic housing and just connects with two wires, there is no problem. You just hook up the wires backwards. If the part has a metal housing, you must separate the housing from any metal parts of the car. This is known as isolation, and may call for placing rubber between the case and the car body or attaching the part using plastic bolts.

If you happen to have a 6 volt negative-ground car, you can often convert it to 12 volts without too much trouble. You will need to use the alternator and voltage regulator from a newer car. Then you can add a second voltage regulator made just for these conversions. This regulator will produce 6 volts to supply the original electronics on the car. Meanwhile, you can use the 12 volt system to run a modern ignition, stereo, or other new parts.

If too much current flows through a wire, the wire is likely to overheat. This can cause anything from melted insulation to smoke and fire. There are three steps to keep the wire from overheating. Step one is to choose the right-sized wire for the amount of current you can send through it. Wire size is measured in gauge, with a smaller number meaning a thicker wire. A 16-gauge wire can safely carry 7.5 amps. A 14-gauge wire raises the limit to 15 amps, while a 12-gauge wire can carry 20. Ten-gauge wire can accommodate 30 amps, which is enough for most parts you will find on a car except starters, alternators, and heavy-duty stereo amplifiers.

Step two is to protect the wire from short circuits. Short circuits occur when a bare-metal wire or terminal comes in contact with the metal of the engine block, chassis, or car body. In this case, electricity takes a shortcut through this contact, and often bypasses the part that the wire was supposed to power in the first place. Short circuits can draw a lot of current and build up tremendous heat if they are not stopped. You can cut down on the odds of a short circuit happening by making sure wires will not come into contact with things that will burn, cut, or rub against them. Also, be sure all connections are insulated well. The twist-on caps used in home wiring are completely unsuitable for use in cars. You'll need to use careful soldering or good crimp connectors to join wiring. Insulate connections with good quality electrical tape, heat-shrink tubing, or liquid rubber insulator.

Step three is to make sure the circuit has some means of limiting the current it carries. If a short circuit occurs, this will shut off the current through the wire. The most common method is a fuse, a block with a length of wire that melts if you send too much current through it. Fuses come in several styles and are rated by how many amps they can carry without blowing and shutting off the circuit. A more sophisticated device is a circuit breaker, a mechanical switch that turns itself off when the current gets too high. Circuit breakers may either require pushing a button to reset them, or reset themselves after a brief period of time.

Whether you use a fuse or a circuit breaker, you should install it as close to the start of the wire as possible. For example, if you need to power something straight from the battery, the fuse should be within a few inches of the positive terminal of the battery. The shorter the length of wire not protected by a fuse, the less room there is for things to go wrong.

Electrical Tools

Working with the electrical system often requires tools for connecting wires and tools for testing the circuitry. A wire cutter and a wire stripper, or a tool that combines both, are essential. Often, you will find yourself wanting to connect a wire to a terminal (a connector at the end of a wire that either works as a plug or is connected to a bolt) or join two wires. Often, the most reliable way to do this is by crimping the wire to the terminal or connector.

Crimp-type connectors may look as if they can be attached by squeezing them with a pliers, but you will need a quality crimping tool to make the connector stay on (not to mention a connector that's of good quality to begin with). A sheet-metal crimping tool can be bought for a few dollars, but the more expensive style that resembles a pliers is money well spent. Ratcheting crimpers can work well if the wire and connector are the right size; the pliers-type tools can work a bit better if you're trying to make something not quite the right size work.

Some electrical work calls for soldering instead of crimping. Soldering requires great care, both to produce a good joint and to avoid burning yourself with the hot soldering iron. To use a soldering iron, connect the wires together and heat them by putting the tip of the iron near the joint. Touch the solder to the spot where the wires join, without touching it to the iron. The heat will flow through the wires and melt the solder. Touching the solder directly to the iron without warming the parts to be joined can make for a solder joint that breaks too easily. Good soldering technique requires practice, but is less dependent on having expensive, top-quality equipment than crimping.

Any place where two wires are joined needs adequate insulation. Joining two wires with no insulation can mean blowing fuses, replacing batteries, or sprinting inside to grab a fire extinguisher to put out a burning dashboard. (Yes, I'm speaking from personal experience on that last one, after not having checked the wiring on a newly acquired project car carefully enough before I turned on the headlights.) Crimp connectors are often already insulated, although many professionals prefer to use uninsulated crimp connectors. A soldered joint can be insulated by wrapping it with electrical tape or its liquid equivalent, but a better method to insulate a soldered or crimped joint is heat-shrink tubing. Slipped over the wires before connecting them together, this tubing covers the joint and shrinks to form a tight fit when heated with a heat gun. In a pinch, you can use a cigarette lighter.

One of the most versatile test tools is a multimeter. This can measure voltage, current, and resistance. With a multimeter, you can make sure a part is getting power, check if a fuse is blown, examine fuel injectors to see if they are burnt out, or look for short circuits.

There are two other test tools that are cheaper and simpler to use than a multimeter. One is a continuity checker. This will have two probes. Clip one to each end of a device. If current can flow between the two probes, a light on the continuity checker will glow. The other is a voltage tester or test light. This tool consists of a probe, a length of wire with an alligator clip, and a light bulb, which is often built into the probe handle. To use the test light, connect the clip to a grounded, bare-metal surface like the engine block. Touch the probe to an electrical terminal. If the terminal is supplied with 12 volts, the light will glow. A variation of a test light called a noid light is a fast blinking design that can check for ignition or fuel injector pulses.

Connecting Electrical Parts

Some electrical parts are easy to wire up. Many devices, like ignition amplifiers, will already come with all the wires they need and directions that tell you where to connect the wires. All you have to do is attach the right connectors to the ends of the wires, connect them where they need to go, and use a few zip ties to hold the wires in place so they will not come in contact with things that might damage them, like hot exhaust systems or spinning fans.

Gauges and similar devices are one step up when it comes to complexity. These can get their power from a wire on the fuse box. For example, many aftermarket gauges have a wire to power a light in the gauge. You can connect this light to the fuse that sends power to the instrument panel lights. If the gauge needs power, you can connect that to the fuse that powers your original gauges.

At other times, you may need a completely new fuse for a part. This usually happens when you're installing a type of device the car never left the factory with. First, you will need to find a point in the car's wiring that is on when you need it and off when you need the part to be off. For example, you might want to equip a classic muscle car with an electric fuel pump. This pump needs to be on when the ignition key is in the Run and Start positions, and off at all other times. Using your test light or a wiring diagram, you can find a point in the wiring that has power at these times and is able to carry enough current to power the new device (it's not going to do much good if your fuel pump needs 10 amps and you're getting power from a circuit protected by a 5 amp fuse). One good spot to look is where the wiring runs from the ignition key to the fuse box. Once you find a stretch of wire, splice in a new piece of wire leading to a fuse. Sometimes you can find an empty space in the fuse box and stick a new fuse in there. At other times, you may need to use a fuse holder.

Other installations use relays. A relay is an electromechanical switch that turns on when you send a current to its coil. The switched side is called the contacts. Use a relay for devices that draw so much current that you do not want to send all the current through the switch you use to turn on the device. For example, suppose that in the previous example you also wanted to use an oil pressure switch that turned off the fuel pump if the engine lost oil pressure (which will often shut down the fuel pump if the engine stops running as well—a useful safety feature in the event of a crash, where you don't want the fuel pump to keep running and spray gasoline everywhere).

In this case, you would find the same point that you connected the fuel pump in the previous example. However, instead of the fuel pump, you would wire this point to the switch in a relay, and wire the relay so the oil pressure switch can turn it on and off. Then run a wire from the relay to the fuel pump.

In a negative-ground car, when you send power to a device, you should send the power wire to the positive terminal on the part, and connect the negative terminal to ground by running a short length of wire to the nearest metal part of the body, frame, or engine. Connect the ground wire with a ring terminal and bolt or screw, and sand the paint off that spot (if there is any) so as to get a good connection. Covering the spot with a spray-on plastic, paint, or sealant meant for rust prevention will make this connection last longer. Remember to make the ground wire thick enough to be safe, and don't ground things to very thin sheet metal or metal that isn't well connected to the rest of the chassis. If you have 15 amps flowing to your fuel pump, there are 15 amps flowing away from it, too. Note that precision electronics, like fuel injection controllers and gauges, are a bit pickier about their grounds—they work best if you ground them to the engine or battery directly, not to the body.

You can use a common wire to carry power to several devices, as long as they all need to turn on and off at the same time or have switches of some sort at the devices themselves. In this case, make sure the wire and the fuse are able to carry the total current drawn by all the objects you are powering. Ground each device and wire all their positive terminals together. This is known as parallel installation.

The opposite is series installation, where you run a wire to the positive terminal on the first part, wire its negative terminal to the positive terminal on the next part, wire the negative terminal on that part to the positive terminal on the next one, and so on, until you come to a part that is grounded. A fuse, for example, is installed in series with the part it protects. A switch installed in series with a part will turn it on and off. Except for fuses, circuit breakers, and various sorts of switches, you should not wire up parts in series unless the directions that come with the part tell you to or you really know what you are doing and why.

Timing Lights

A timing light is used for adjusting the performance of the ignition. They are sometimes called timing guns due to their pistol-like shape, or timing strobes because of their rapid blinking. To use one, connect its power wires to the battery, clamp the pickup on the number-one spark plug wire, and aim it at the large pulley on the front of the crankshaft while the engine is running. The timing light will light up the pulley with a strobe light that will make the markings on the pulley appear to stand still. The markings will indicate when the number-one cylinder is firing.

There are a couple of features to look at when buying a timing light. The first is an inductive pickup. This works by simply clamping it around a spark plug wire. There are a few older or very cheap timing lights that have to physically touch the wire that is carrying current, either by putting a length of spring on the end of the wire or by puncturing the insulation. This is not the safest approach, to put it mildly. The second feature is to make sure the light is bright enough to see in broad daylight; some of the cheapest models may not be visible except in a dark garage. One feature found on higher-priced lights is an adjustable advance knob. This allows you to see the timing settings when the engine is running at full throttle as well as at idle, since the timing moves with engine RPM and often moves beyond the scale of timing marks on the engine. Note that some cheaper dial backlights may be designed to work only with a distributor and may have some issues with certain types of distributorless ignitions. (More about what these ignition types are later.)

Fabricating Tools

Some enthusiasts practically own their own machine shops, with a lathe and milling machine in their basement. This is a bit extreme, but you may still want a few small metalworking tools. A power drill and a good set of drill bits are often essential. When it comes to drill bits, you get what you pay for, as cheap drill bits are made of cheaper metal and will not stay as sharp as long. Using a center punch and a hammer will make drilling accurate holes much easier.

A Dremel or other high-speed rotary tool is amazingly versatile. With the right attachments, it can sand, polish, or cut through metal, but is only suitable for small jobs and precision work. Cordless rotary tools seldom have enough charge for major automotive jobs, so you will not want one with batteries. Chisels and hacksaws are also very useful. A power sander, angle grinder, or jigsaw can be useful in some situations, but you can hold off on buying these until you have a project that calls for one.

It's worth spending extra for a top-quality drill.

A rotary tool can grind, cut, or polish.

A reciprocating saw is great for big cutting tasks.

Tube Tools

If you are making fuel lines or brake lines, you will need specialized tools. Although you can cut the tubing with a hacksaw, a tubing cutter will produce a good, clean, perpendicular cut, and do so in considerably less time. This will not only make the tube easier to cut, but it makes installing fittings easier.

Hard tubing typically uses flare-type fittings. These screw-in fittings slip over the tube, and then the end is expanded with a flaring tool. Flaring tools come in several varieties, and it is important to use the right tool for your fittings and the application. Most household water tubing uses a 45-degree single flare. This works well for pipes that sit still, but in high-pressure brake lines, the edge can form cracks that cause the tube to split. For brake use, you will need to use a 45-degree double flare, which folds the edge of the tube back in on itself. A double-flare tool can make single flares, but a single-flare tool will need additional parts to make a double flare. Double flares are also sometimes called inverted flares.

Some high-end street cars and many race cars use AN tubing. Developed for the military (AN stands for Army-Navy), this system uses a 37-degree single flare, and uses other changes to the tube material and fitting designs to avoid cracking. These require a specific AN flaring tool. If you are working with AN fittings, you might also want to buy a set of stubby AN wrenches designed to tighten these fittings without overtightening or scratching the aluminum finish.

AN flares work with AN hoses. Many AN hoses have a braided stainless-steel outer layer for extra safety. These hoses can slip over normal hose barbs and attach with hose clamps, but the best way to attach them to tube or pipe threads is with their own specialized hose ends. These fittings typically have two pieces, clamping the hose between an inner nipple and an outer socket. Braided steel lines require a few tricks to attach them to their hose ends, and can make you pay for mistakes with your own blood. Granted, it's not much blood, but the loose ends of the steel braid will cut careless fingers.

The best way to attach a hose end begins with a careful cut. To keep the braid from unraveling, wrap the hose with electrical tape where you want to cut it. While you can cut the hose with a fine-toothed hacksaw, an abrasive cut-off wheel or a braided hose cutter often produces a neater cut. After cutting the hose, leave the tape in place and thread a socket onto the end of the hose, holding the hose still and turning the socket counterclockwise. Once the socket is firmly on the hose, you may peel off the electrical tape and tighten the socket the rest of the way. Then lubricate the fitting body with motor oil and thread it into the socket.

A budget tip—there's another type of hose end called JIC, intended for industrial use (the name JIC stands for Joint Industry Council). JIC hoses are pretty much interchangeable with AN on a car, but cost significantly less.

Another flare that you will sometimes see is the "metric bubble" flare. Strangely, this "metric" design often gets combined with inch-sized tubing and fittings. Tools to make these flares and fittings to work with them are hard to find. In some cases, you may be able to use regular double flares and adapter fittings.

Hard line does not always require a flare to safely secure its ends. Another alternative is to use compression fittings, which have a ring called an "olive" that slips over the line. Tightening a compression fitting tightens this ring against the line for a firm seal. Compression fittings are suitable for fuel, oil, or water lines.

While you can simply stick the end of a tube into a hose and secure it with a hose clamp, this sort of connection can be prone to leaking. One solution is to flare the end of the tube and attach a fitting to connect it to a hose barb, but this doesn't make it easy to get the hose on. A more elegant solution is to use a beading tool to create an enlarged spot just below the end of the tube, allowing the hose to seal against the tube. Beading tools don't come cheap, though.

A tube bender can be a useful tool, but it is not always necessary for working with hard line. Thin metal tubes can be bent by carefully twisting them around a baseball bat, length of water pipe, or other solid, round surface. Take care not to bend it hard enough to crimp the tubing. A tubing bender can make life much easier with large-diameter tubing such as 3/8 inch fuel lines.

Torches and Welding

You normally do not need a torch when you start working on cars. Many people put together excellent cars without any welding. However, if you want to make custom headers, repair rust, or make your own parts from sheet metal, you will want a welding torch. In fact, you may want a couple of different kinds of torches.

Gas welding with an oxyacetylene torch requires a considerable amount of practice. You probably won't want a very large gas torch unless you already know how to weld with one or plan to do a lot of cutting. However, the oxyacetylene torch's smaller relative, the MAPP gas torch, is quite handy for cutting metal and loosening stuck bolts. These torches can be bought for as little as $50, but the disposable oxygen bottles can get pricey. Inexpensive single-bottle torches do not put out enough heat for cutting or welding, but they can be useful if you want to solder pipes.

Arc welding is the most popular sort of welding today. This uses a high-voltage electric current to create enough heat to melt two pieces of metal and join them together. Arc welding often requires two other components: filler material and a shielding gas. The filler material is extra metal that melts to fill in gaps. Shielding gas is an inert gas that keeps the hot metal from reacting with the oxygen in the air. This gas may be applied through a nozzle, or it may be produced by a flux applied to the filler material that vaporizes when the metal melts.

The most familiar sort of arc welding is stick welding. This uses a power supply to send current through a flux-coated welding rod. In the hands of an experienced welder, stick welding can produce the best welds to join thick steel plate or iron tubing, making it useful for building turbo manifolds. The downsides are that it takes a lot of practice to reach that "experienced welder" state, and even experienced welders prefer other methods for the thin sheet metal of a car body.

One of the easiest sorts of welding to learn is MIG welding. This stands for metal inert gas. An MIG torch tip supplies both high-pressure gas and a steady flow of filler material fed from a spool of wire. With a correctly adjusted machine and a few hours of practice, it is not hard to get the basic idea of how to use one. Some MIG units can also use flux-cored wire, which may not produce as clean a weld but can be somewhat easier to use in the wind. You can find functional MIG welders for as little as $200, but professional welders prefer the more expensive units from Miller or Lincoln Electric.

The TIG, or tungsten inert gas, welder is popular for precision work. A TIG torch supplies gas like a MIG, but any filler must be supplied from a handheld piece of wire. Some TIG welding leaves out the filler. TIG welding delivers its heat at a very small point. This can produce some of the smallest and best-looking welds short of welding with a laser beam. An expert welder with a TIG torch can weld two Coke cans together without melting through the thin aluminum. Many stick welding units can also be used for TIG welding with the right attachments. Like stick welding, this takes more practice than MIG.

Arc welding produces a blinding flash of light and a considerable amount of ultraviolet. A welding helmet suitable for arc welding covers the entire face, not just the eyes. It is well worth the money to get a helmet with a visor that automatically darkens when you turn on the current, instead of one that you have to flip up to see when the current is off. To be safe from both sparks and weld-induced sunburn, it is best to wear heavy clothing over your entire body while welding, and to avoid wearing nylon or polyester. One last danger is the welding fumes; be sure to do the welding in a well-ventilated area.

Suspension Tools

Often, "bolt-on" suspension parts may need a few tools besides ordinary wrenches. The most well known is a coil spring compressor, which is mandatory if you are changing springs or struts (unless, of course, you're working on a car that doesn't have any coil springs). Serious suspension work may require special tools to remove ball joints or steering parts. One especially handy tool is known as a pickle fork because it looks like an overgrown utensil for serving pickles. This can be used to separate ball joints or tie rods, although if you use this you can plan on having to replace the rubber boots that protect these parts. A good quality pickle fork will be made in one piece; those with the handle and end made separately and welded together can have their welds break.

And More . . .

There are even more types of tools sometimes needed while working on the car. These are usually only needed for internal engine work or other extreme modifications, although some cars may require specialty tools for other purposes. In any case, the less common tools will be something you can wait to buy until you have a job that needs them. In many cases, parts stores will lend you the tool if you leave them with a refundable deposit.

CHAPTER 4

Engine—Intake and Exhaust

Simple Ways to Get More Power and Better Sound out of Your Engine

An engine is a complicated machine with hundreds of parts. However, in this chapter, we're covering what is often the simpler set of engine mods, the kind you can install without taking the engine almost completely apart. These are intake and exhaust mods, and it's possible to get a decent understanding of what you need here without looking too much inside the engine. In this case, simpler mods can start with a simpler understanding. We'll go over the internal workings of the engine and how to modify those parts in the next chapter.

For this chapter, we can think of the engine as an air pump. It pumps in air, mixes it with fuel, lights the mixture on fire, and then pumps out the spent gases. Anything that makes it harder for the engine to pull in air or push out the exhaust reduces power. In this chapter, we're going to focus on how to remove obstacles to this pumping.

This is also a good time to bring up the topic of engine displacement, which is the number used to tell an engine's size. For example, a 5.0 Mustang engine has 5.0 liters of displacement. The displacement is how much air an engine would pump in and out in one cycle if it were a perfectly efficient pump. (We'll get into exactly what a cycle is for the engine in the next chapter.) A larger displacement engine is going to use more fuel (hopefully to make more power) in one cycle than a smaller one.

Intake System

Most cars use a pipe, tube, or scoop to draw in air from outside the engine compartment. The incoming air first flows into an air cleaner that contains a filter to remove dust, bugs, and other things that can shorten the life of the engine. After leaving the air

Get air and fuel into the engine; get horsepower and exhaust out. (*Photo courtesy Edelbrock.*)

cleaner that holds the air filter, the air flows to the throttle, a valve that regulates the amount of air the engine can draw in. On modern fuel-injected engines, the throttle is usually set in a small metal block or plate known as the throttle body. Older engines use a carburetor, a mechanical device that includes one or more throttle valves and a mechanical system for metering (measuring) the amount of fuel required and pouring it into the incoming air. While it is common to have only one carburetor or throttle body, some engines may use two or more.

The throttle body or carburetor is attached to the intake manifold, a structure that distributes air to the cylinders. Most intake manifolds combine a large open space called a plenum with a set of runners, smaller passages that lead to the cylinder head, and from there into the engine. Fuel-injected cars will have fuel injectors at the ends of the runners. The injectors are electronic valves that open to spray fuel into the airstream.

Intake Modification

There are many aftermarket parts designed to make the intake system less restrictive. Heavily modified engines may have every part of the intake system replaced with high-performance pieces in a quest for more power. Some of the parts can be replaced with few trade-offs other than possibly more noise. Other parts may affect drivability, making the engine idle poorly or making the throttle control less precise. These compromises are more likely to happen on cars that use carburetors, but in some cases a fuel-injected car may not be quite as docile if it has aggressive intake modifications.

The simplest part to replace is the air filter. Paper filters will stop a lot of crud, but they create some degree of restriction. There are several aftermarket filters made from cotton gauze or foam rubber that will flow better, although only replacing the filter is

A foam air filter and air cleaner assembly. (*Photo courtesy Edelbrock.*)

not likely to give more than a 1- to 2-hp improvement or so. High-flow filters are also a bit less effective at stopping dirt than the stock paper ones, although not necessarily different enough to cause more engine wear. Never replace an air filter on a street car with wire mesh—not only does this not keep out anything smaller than gravel, but some mesh "air filters" actually are more restrictive than paper. Other improvised filters like nylon stockings are equally bad.

Some modern cars respond well to replacing the air cleaner and air intake tubing. The original setup is often designed as much for quiet operation as for performance. The most common replacements are short ram intakes and cold air intakes. A short ram intake uses a cone-shaped air filter with no air cleaner and locates it on a short piece of pipe under the hood. The goal is to have as little use of pipe as possible for a system that can, in theory, have the least restriction. The problem with a short ram intake is that it draws in hot underhood air, and hot air is less dense than cold air. This means that hot air will have less oxygen per liter of air, and you will not be able to burn as much fuel. Consequently, a short ram intake can sometimes cost you power when the engine is warmed up, but may be useful in drag racing if you have a chance to open the hood and ice down the engine compartment between runs. It also may work well if it happens to put the filter in a relatively cool spot of the engine compartment near an opening to outside air.

Cold air intakes, like their name implies, draw in cold outside air. The extra length may be a minor restriction compared to a short ram intake, but the lower temperature almost always makes up for this. When looking for a cold air intake, make sure the filter is not in a location where it is likely to suck up water if you drive through a puddle. Filling the cylinders with water can completely ruin an engine. A filter mounted behind the grill is likely to be safe, and an engine can suck up a few raindrops without harm. One located inches from the ground or in the fenderwell, with nothing to shield it from water kicked up by the tire, could be trouble if you are not careful when driving through standing water.

A short ram intake replaces a long length of restrictive factory intake tubing.

There are a number of methods to help prevent a cold air intake from becoming a cold water intake. The simplest step is to put a splash shield between the filter and any likely place for water to approach it. A more sophisticated solution is a water bypass valve, which opens in response to water causing a restriction in the intake. AEM makes a bypass valve that they've demonstrated by putting the inlet of the intake in an aquarium while making a dyno pull—the valve didn't let any significant water in. Consider running one of these if your cold air intake puts the filter in a vulnerable location.

It is not difficult to make an educated guess as to how restrictive the piping in an aftermarket intake will be if you do not have any dyno tests of this particular intake. In most cases, an ideal intake will be the same diameter throughout its length and have gentle, flowing bends. Some changes in diameter may not hurt if they are very gradual and flow well into each other. Tight bends, miter bends, and abrupt changes in diameter all create restriction.

Fuel-injected engines need a means of measuring how much air they draw in. Many engines use a mass airflow sensor. There are two common types: vane airflow meters and hot wire sensors. Hot wire sensors measure how much the incoming air cools down a heated wire. These are not normally a major restriction, but if you have increased the diameter of your intake upstream or downstream, the stock unit may limit how much you can gain until you replace it with one that matches the diameter of the rest of your intake system. Vane airflow meters use a spring-loaded vane that sticks out into the airflow. The more air an engine pulls in, the further it pushes the vane. This is more restrictive than a hot wire sensor, and in some cases you may find a kit that allows you to replace your vane airflow meter with a hot wire sensor—or remove it altogether. A few cars use a Karman vortex meter, which is somewhat in between a vane airflow meter and a hot wire design in terms of restriction.

Not all fuel-injected engines use mass airflow sensors. Some use a MAP (manifold absolute pressure) sensor that measures the pressure downstream of the throttle. Pure race systems may dispense with measuring the air at all and use the throttle position to determine how much fuel to add. In either case, you no longer have a sensor that the air has to physically flow through. I've seen a Miata equipped with an aftermarket engine control unit that allowed changing sensor types gain 5 hp when removing the vane airflow meter and using a MAP sensor instead. When you consider that the car only made around 90 hp on a chassis dyno, this gain is even more impressive.

Many carbureted engines have another restriction in the intake system. The air cleaners that many of these engines use force the air to make a very sharp bend as it enters the carburetor. There are a handful of ways to smooth out this transition. One is to run an air cleaner designed with smooth curves, which gives the air a straight shot down into the carburetor. Another is to add a device like a K&N Stub Stack to the existing air cleaner.

The next restriction that can be dealt with in the intake system is the throttle. On a fuel-injected engine, simply replacing the throttle body with a larger one would seem logical. This is true, but if the throttle body is too large, the throttle will behave almost like an on/off switch. It is possible to get around this to some extent by using a progressive throttle body, which has two or more throttles that open one after another. If you haven't done major internal work to the engine, a new throttle body should not be much larger than the one it replaces.

There are a number of ways of getting a larger throttle body. If the throttle body is made of metal (most are, but sometimes you'll see plastic ones), it is sometimes possible to bore it out by using a cutting tool to enlarge the hole, and cut a larger throttle plate from sheet metal. The junkyard approach is to adapt a larger throttle body off a different

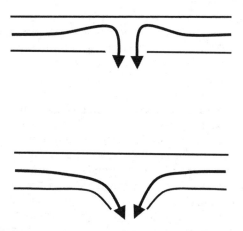

Cross-section of two air cleaners. The one at the top forces the air to make a tight turn; the one at the bottom allows for a gentler bend and will flow better.

An aftermarket throttle body can let in more air than the stock one. (*Photo courtesy Edelbrock.*)

car. Sometimes the throttle body may bolt directly on, particularly if it is from the same manufacturer. At other times, it may be possible to make an adapter. If you need a monstrous throttle body, companies like Accufab, Edelbrock, or BBK make oversized throttle bodies from all new parts. When installing a new throttle body, make sure that the openings upstream and downstream are large enough to accommodate it. Bolting on an 85mm throttle body won't help too much if it's on a manifold where the hole for the throttle is only 65mm.

Carburetor sizing is more critical. A carburetor relies on creating a partial vacuum to operate, and this requires a bit of restriction. A carburetor that is too large will not be able to mix fuel accurately at low RPM. Progressive throttle systems are especially important in carburetors, since a four-barrel carburetor with four throttles and a progressive opening system can run at low RPM with two of its throttles shut and only two of its barrels active at times when using all four barrels would not generate enough vacuum.

Progressive opening carburetors are divided into two types: those where the secondary throttle plates are opened mechanically when the gas pedal is pushed down all the way (mechanical secondaries), and those where the secondary throttle plates are opened in response to the airflow demand becoming too much for the primaries (vacuum secondaries or air valve secondaries). Many prefer vacuum or air valve secondaries for street use, since they are somewhat more forgiving when it comes to choosing size. Opening all four barrels instantly with mechanical secondaries can cause an engine to bog or hesitate if the engine is not turning fast enough already, especially if the carburetor is too large for the engine or not tuned properly. Vacuum secondaries do not open until they are needed, reducing this problem.

The Edelbrock 1405 is a four-barrel carburetor with air valve secondaries.
(*Photo courtesy Edelbrock.*)

HOW TO SELECT THE RIGHT SIZE CARBURETOR

Most American carburetors are sold not by any size measurements, but by their airflow rating. Normally, this is measured in cubic feet per minute. A carburetor should be rated to flow as much air as the engine will need, but one that is larger than necessary may not be able to operate correctly at low RPM. The amount of air your engine needs is relatively simple to calculate. If you are measuring your engine in cubic inches, the formula is like this:

Airflow in cubic feet per minute = Engine size in cubic inches × Maximum engine RPM × Volumetric efficiency percentage / 3,456

If you don't know the volumetric efficiency, use 85 percent for an older and relatively stock engine, up to 95 percent for a fairly heavily modified racing engine or stock modern performance engine. A very highly developed engine can even go over 100 percent.

For those wanting to put a carburetor sized in English units on a metric engine:

Airflow in cubic feet per minute = Engine size in cubic inches × Maximum engine RPM × Volumetric efficiency percentage / 56.7

So, this number will tell you what carburetor size to buy, right? Not yet. This formula tells you how much air your engine needs, not how large your carburetor needs to be. Carburetors are rated at a pressure drop of 1.5 inches of mercury. You don't want to see that much pressure drop across the carburetor on a real engine at full throttle. That much restriction will

Two-barrel carburetor flow numbers are measured with twice the pressure drop, making their flow numbers appear larger.

hurt power. To get a carburetor that won't hurt power, multiply this number by 1.7 and you'll have a number closer to the size you need.

If the manifold has a single plenum fed from all the carburetors, then you can use one carburetor that flows this much or a combination of carburetors that add up to this flow level. If your manifold has a divided plenum, you can use a somewhat larger carburetor. Independent runner intakes, where each cylinder draws fuel from only one barrel of a carburetor and the manifold has no plenum, are not sized using this method. Most companies that sell independent runner intakes will be able to recommend a suitable set of carburetors for your engine.

Curiously, two-barrel carburetors use a different standard of measuring their flow rates. Flow rates for two-barrel carburetors are measured with twice the pressure drop used when measuring four-barrel models. To convert the flow rating of a two-barrel carburetor to the standard measurement, multiply it by 0.7072.

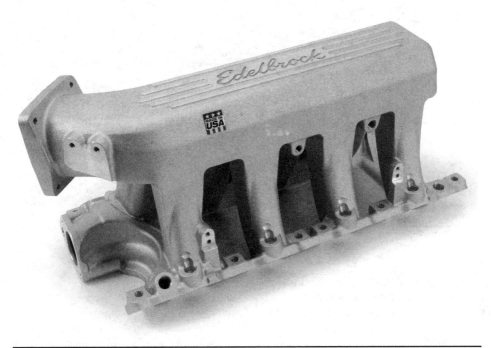

Most EFI intake manifolds use a simple plenum-runner design. This one from Edelbrock is for Ford V8s. (*Photo courtesy Edelbrock.*)

Replacement intake manifolds are often necessary if installing a different carburetor, and may give you a performance gain too. While factory fuel-injected manifolds often have fewer compromises than carbureted manifolds, an injected engine with significant internal modifications may also benefit from a manifold upgrade. Most intake manifolds have two distinct parts: a plenum immediately downstream of the throttle body, and separate runners branching off to each cylinder. As a general rule, making the runners longer and thinner can boost performance at low RPM, while short, wide runners are less restrictive, enabling an engine to make more power at high RPM. Aftermarket intake manifolds also frequently use a larger plenum, which can reduce restriction at high RPM but can also hurt throttle response if you go too big.

On carbureted engines, there is another consideration. The fuel is added at the carburetor, so the manifold contains a mixture of fuel and air. Some designs can cause the fuel to separate from the air in places, meaning that some cylinders get more fuel and others get less. Manifolds that make the air flow uphill are especially likely to have this problem. To complicate matters further, the way the manifold works can affect how well the carburetor mixes fuel. These manifolds are often divided into dual-plane types, which have a split plenum, and single-plane types, which have a single open plenum. A dual-plane manifold will send fuel and air from one side of the carburetor to half the cylinders, and the air and fuel from the other side of the carburetor goes to the other half of the cylinders. In general, a dual-plane intake manifold is likely to work best at low RPM, and a single-plane manifold would work better on a high RPM engine.

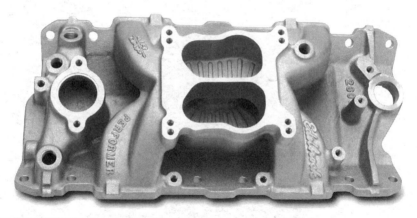

Dual-plane intakes are good for mild street V8s. (*Photo courtesy Edelbrock.*)

Unfortunately, airflow through a manifold can be difficult to predict, making design of a manifold for a carbureted engine as much of a black art as a science. It's hard to tell whether one manifold will perform better than another of the same basic design by looking. If possible, find a report where someone has tested the manifold you are looking to buy on a dyno. Also, most manufacturers will rate their manifold for the

A single-plane intake for high RPM racing. (*Photo courtesy Edelbrock.*)

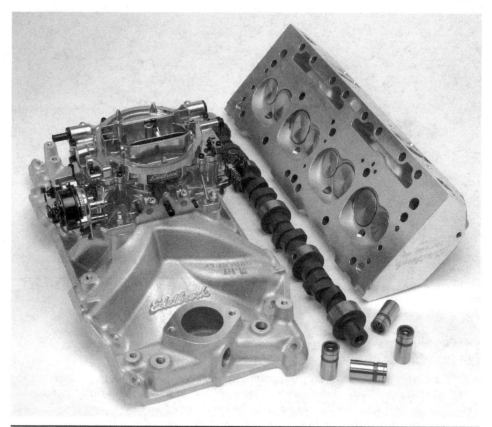

This package includes an intake manifold, carburetor, camshaft, and cylinder heads, all intended to work together. (*Photo courtesy Edelbrock.*)

RPM range where it will be most effective. For best results, you should select all the performance parts to work well in the same RPM range. In fact, many vendors have taken to selling intake manifolds and parts for internal mods in a single, matched package.

Exhaust System

The engine will push spent gas out the exhaust system, and the less work it takes to push the exhaust out, the less power is wasted. There are two types of devices the exhaust might exit the engine into: an exhaust manifold or a header. An exhaust manifold is usually made of cast iron, while headers are made from steel tubing and are often designed to flow better than a manifold. Either one contains a set of passages to direct the air into an exhaust downpipe. The air then passes through one or more catalytic converters (or cats for short) to remove some of the more harmful pollutants, and then goes into the exhaust pipe, where the exhaust is quieted by one or more mufflers, and possibly a resonator (basically a smaller and less restrictive muffler to fine tune the way the exhaust sounds) before leaving the exhaust tip.

A complete race exhaust for a V8, with headers and an X pipe. (*Photo courtesy Edelbrock.*)

Exhaust Modification

Pushing the exhaust out of the engine requires power. Removing restriction from the exhaust can reduce the power needed, leaving more power to turn the wheels. Many heavily modified cars have each and every one of the parts of the exhaust system replaced in a quest for more horsepower. Improving the exhaust will decrease back pressure, the pressure of the exhaust where it leaves the engine.

Sometimes a less restrictive exhaust can flow too well, but only for low RPM. This usually occurs on engines tuned to run at very high RPM. In this case, at low RPM, the gas flowing out the exhaust valve can actually suck the incoming air and fuel out the exhaust valve at low speeds. As the engine speed increases, this no longer happens because back pressure rises while the valve is open for less time. So on some engines, less back pressure will result in less torque at low RPM. Other engines do not have this problem. In virtually all cases, less restriction will mean more peak horsepower. Whether it may hurt torque at low RPM is very difficult to predict unless you already have a dyno test of the exhaust system you want on the same engine you have, with the same mods you plan to use.

The most common exhaust mods simply replace the exhaust piping and mufflers. The muffler is one of the largest restrictions in the exhaust. If the rest of the pipes are acceptable or you have a very low budget, you might even want to replace the muffler on its own. There are many ways to design a muffler, but most performance mufflers on the market today can be divided into three categories: chambered, turbo, and straight-through or glasspack. Mufflers are usually designed to work more for a given horsepower level or flow level than for a particular sort of engine, so a muffler that performs well on a 300-hp American V8 will probably also give good performance on a 300-hp turbocharged Japanese inline four, and vice versa. The sound may be different, but the performance is likely to be the same.

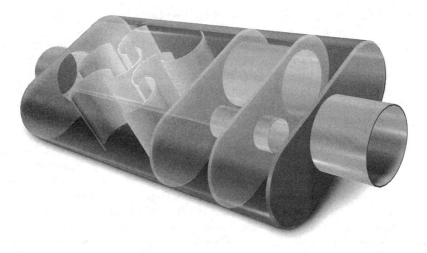

Cutaway of a chambered muffler. (*Photo courtesy Edelbrock.*)

The straight-through muffler is a simple enough design that it is often possible to tell a good one from a poorly designed one simply by looking. A free-flowing glasspack muffler will have a tube through the center that is the same diameter as the inlet and outlet pipes. Some glasspacks have a narrow tube that can actually make it perform worse than the stock muffler. For minimum restriction, this tube should not have louvers facing into the inside of the tube. Spirals, ridges, and other features will also create more restriction. The tube should have holes or outward-facing louvers instead. This makes for good flow, but also makes for a rather noisy muffler.

Glasspacks are not the most durable mufflers. The fiberglass in a straight-through muffler may eventually break and fall out, making things considerably louder and

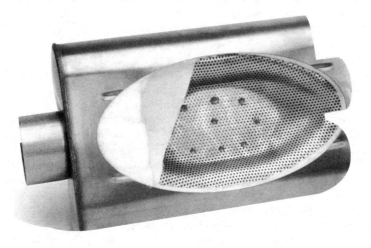

Cutaway glasspack muffler. (*Photo courtesy Edelbrock.*)

possibly requiring buying a new muffler. Ceramic fibers instead of glass, and stainless steel instead of galvanized steel, offer a longer life. Most of the other features, such as fancy tips or polished cases, are more about looks than performance. Buying a $120 straight-through muffler may not get any better performance than a $20 version.

Although the name suggests that a turbo muffler is designed to work only with a turbocharged engine, turbo mufflers are simply mufflers that resemble the muffler originally used on the '60s-era Chevy Corvair Turbo. Most turbo mufflers have three cores in the center similar to a glasspack, combined with open chambers at the end to reflect sound in such a way that it does not escape the exhaust system. Chambered mufflers rely almost entirely on baffles to reflect the sound. Some of these mufflers have no glass fibers to blow out, and can last much longer than glasspacks.

The quality of aftermarket mufflers varies. Some will flow well while keeping the engine quiet. Others flow well but are noisy. Some "performance" mufflers on the market actually flow worse than the originals, and original mufflers on many of today's high-performance cars are often designed well enough that any designer wanting to improve performance without making things louder will have a difficult time.

The best way to tell if a muffler will work for your car is to find an actual dyno test of an engine set up the same way as yours to see how much power it gains over stock. In many cases, you will not be able to find a test this specific. The second best way to tell if a muffler will work is to use flow testing data. A common rule of thumb is that you will want a muffler that will flow 2.2 cubic feet per minute or more for every horsepower your engine would make with no mufflers. If you use dual exhausts, you can add the flow of the two mufflers to get enough flow. Alternatively, you can select a muffler that produces very little power loss on an engine that makes as much power as you hope to make. For how a muffler sounds, you will need to visit races and car shows and hear it for yourself to see if it is the sound that you want. The Internet has made researching sounds easier, as a lot of video hosting sites like YouTube have clips of how an engine sounds with a particular set of mufflers.

Although you can have a muffler shop install a new muffler on your existing exhaust pipes, many aftermarket kits replace all or part of the piping as well as the muffler or mufflers. The most common sort of aftermarket exhaust is the cat-back, which replaces all the exhaust pipes downstream of the catalytic converter. Sometimes you may see axle-back systems, which only replace the pipes behind the rear axle; header-back systems, which remove the catalytic converter entirely; or turbo-back systems that do the same for turbocharged cars. A header-back or turbo-back exhaust is unlikely to be street legal unless you install it on a car built before 1975.

Exhausts are also classified as single exhausts or dual exhausts. A true dual exhaust has two separate sets of pipes, one handling exhaust from half the cylinders and the other handling the exhaust from the other half. On a V-type engine, this almost always means each side of the engine has its own exhaust system. Some inline six engines use split manifolds that direct the exhaust from the front and rear cylinders into separate pipes. Most of the time, four-cylinder engines do not benefit from true dual exhaust systems. Some exhausts may start with one pipe that splits in two to accommodate two mufflers, which can flow better than one single muffler. A few exhausts pretend to be dual exhausts by splitting downstream of the muffler.

There are several important features to look for when selecting an aftermarket exhaust kit. Besides making sure it comes with a good muffler, the next most important factor is the exhaust pipe size. Usually, bigger is better, although in some cases a too big exhaust may make a little less power than one that's sized perfectly. As a general rule, a naturally aspirated four-cylinder engine or six-cylinder motors in the 200-hp range will almost never need an exhaust larger than 2.5 inches in diameter, while a V8 can use dual 2.5-inch exhausts for up to 400 hp. A single 3 inch is a good candidate for a six cylinder or V8 in the low 300-hp range, while a dual 3-inch exhaust can, as you might expect, work with roughly double that power. If you're looking for a commercial, ready to run exhaust, though, you might also want to look at dyno graphs of how it compares to competing models, preferably from a neutral source. Real performance beats theory any day.

On a turbocharged engine, it's less important to try to find the perfect tailpipe size and better to just go with the biggest diameter that fits your chassis and budget. The turbocharger tends to break up the individual exhaust pulses and results in a more continuous stream of exhaust flow, so you don't have to deal with the same level of weird pressure waves or other effects that can cause pipe diameter to affect horsepower. A massive exhaust pipe is unlikely to reduce power on a turbocharged engine. In fact, the less restriction in the exhaust after the turbo, the faster the turbo will spool up. The only way an oversized exhaust is likely to slow down a turbo engine is if it weighs too much or drags on the ground.

There are several other features to look for on an exhaust kit. The most important one from a performance standpoint is mandrel bent tubing. This tubing has smooth bends that theoretically are less restrictive than the cheaper crush bends. Most of the other features are for improved durability and better fitment. An exhaust should mount from rubber mounting isolators like the original design to reduce rattle. A flexible coupling, preferably made of braided stainless steel, between the catalytic converter and the rest of the exhaust will prevent engine vibrations from damaging the exhaust. Unless the exhaust has a flexible part between the engine and the first point where it attaches to the car body, the vibrations from the engines will put stress on the flanges and can burn gaskets if the vibration pries two flanges apart. This is especially a problem on sideways (transverse) mounted front-wheel-drive engines, which rock back to front instead of side to side as they vibrate.

There are also important differences in an exhaust pipe's construction. Some exhaust systems use slip fittings, which can be difficult to fasten and prone to leaks, while others use flanges and bolts. High-end systems often have V-band clamps that allow for easy assembly and disassembly. Tubing may be made from mild steel, stainless steel, or exotic materials like Inconel or titanium. Mild steel is the cheapest, but also most prone to rusting. Titanium offers some corrosion protection plus light weight. Stainless steel is one of the most durable choices, and can be polished for an absolutely beautiful finish.

If your car has dual exhausts on a V8, connecting the two pipes together will usually improve performance. This mostly applies to V8s, which will put out more exhaust on one side on one crank revolution, and then put out more on the other side on the next turn. A crossover between the dual exhausts will even things out. An H-pipe, which is simply a tube running from one exhaust pipe to the other, is a common way of doing this. A recent development is the X-pipe, where the two pipes come together in an X shape. X-pipes appear to be less restrictive and provide more power in many cases. This is less important on six-cylinder engines, which have more even flow from side to side.

The exhaust tip is almost always a purely cosmetic choice, except for some extremely crimped designs that may be a little restrictive. Some owners may opt for a large, polished tip, while others may go with an unobtrusive turn-down tip to hide the size of

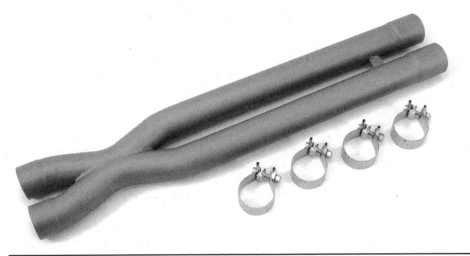

An X-pipe. (*Photo courtesy Edelbrock.*)

their exhaust from potential opponents. Mufflers that mount at the back of the car usually have the tip built in.

You can have a local muffler shop build a performance exhaust. Often, even a cheap exhaust can deliver performance comparable to a good cat-back if you use the right muffler and appropriately sized pipes. The catch is that a $200 exhaust will probably not fit as well, rattle more, look cheaper, and sound less refined than a top-quality aftermarket exhaust. To get these improvements, you will likely need to pay more for a custom exhaust than a mass-produced one of equal quality. It depends on whether you want as much performance for your money as possible or are also concerned about how your exhaust looks and fits.

As a general rule, modifying the exhaust downstream of the catalytic converter is not going to violate any emissions laws. However, it may violate local noise laws. It is also a good way to attract attention, not always of the sort you want. Check your local laws before installing a louder performance exhaust. Some laws set a maximum noise level measured in decibels. More tyrannical areas may forbid any exhaust work that makes the car louder than it was originally.

Modern catalytic converters work very well and seldom reduce power on a typical street engine. The most common exception today is that some cars have what is known as a close-coupled catalytic converter that is a part of the exhaust manifold. Installing a header and a cat mounted under the floor can improve power in these cases. The other common case is that older cars from the '70s and early '80s used highly restrictive pellet-type converters. These are like blowing the exhaust through a bunch of aquarium gravel. Swapping these for modern honeycomb designs can free up a considerable amount of power.

There are strict laws about replacing catalytic converters. Federal laws generally require a catalytic converter on cars built after 1975 to drive on the street, with a few exceptions. Exhaust shops are usually not permitted to remove a properly functioning one even to replace it with one that keeps the exhaust just as clean. On the other hand, it is certainly legal to replace one that has become clogged or otherwise damaged. If you swap out an old pellet-type converter for a high-performance honeycomb one in your own garage, you probably won't have to deal with the same sorts of legal issues that a shop faces. Of course, if your car is built for racing and not driven on public roads, virtually anything goes.

If you want to measure how restrictive your exhaust system is, you can install a temporary back-pressure gauge. To do this, drill a small hole (usually smaller than a quarter of an inch) in the exhaust in front of the catalytic converter. Connect a few feet of copper tubing to this hole, either with threads or by brazing it in with a propane torch. The copper tubing is needed to give the exhaust room to cool off. Connect a rubber tube to the other end of the copper tube and run it to a pressure gauge sitting somewhere in the passenger compartment. Find a stretch of pavement (preferably at a dragstrip) where you can safely floor it in second gear and run the engine up to redline. Ideally, you will want to have a friend in the passenger seat watching the gauge, but you can do this yourself. Take note of the highest pressure reading. Anything more than 1 or 2 pounds per square inch (psi) is probably robbing you of horsepower. Below those levels, it's tough to tell if you're losing horsepower from just the pressure.

For the adventurous, you can piece together the parts needed for a back-pressure gauge with parts from a hardware store. However, if you do not want to try to piece one together, many well-stocked tool sources sell complete kits that include everything you need except the drill.

Cast iron exhaust manifolds are usually designed for low cost and decades of durability rather than maximum performance. Replacing them with tubular headers is a popular mod. A normal set of tubular headers consists of individual pipes known as primary tubes leaving each exhaust port before coming together in a collector pipe that connects to the exhaust pipe or catalytic converter.

Tube headers can provide a power increase in three different ways. First, the tubes are often less restrictive than the tight bends in a cast iron manifold. Second, the exhaust

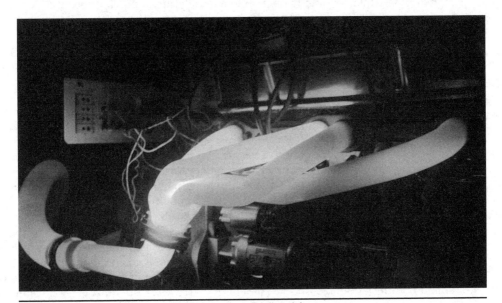

Headers glowing in a dyno test. (*Photo courtesy Edelbrock.*)

pulses create pressure waves that bounce back when they reach the collector, and a header of the proper length can take advantage of this by having a low-pressure wave reach the exhaust port just when the exhaust valve is opening. Third, the pulses interact with each other at the collector to pull each other along in an effect known as scavenging.

In theory, a set of equal length headers would provide the best scavenging and the best power. In practice, more important considerations include putting as few bends in the tubing as possible (and avoiding extremely tight bends altogether), selecting the correct diameter tubing, and getting each tube fairly close to the length needed for best performance at the desired RPM. As a general rule, the longer the tubes, the lower the RPM where the header will be most effective. The exceptions are "shorty" headers designed to fit into cramped engine compartments, which may offer less restriction but are usually neither equal length nor tuned to take advantage of pressure wave effects. The number of bends and diameter of the pipes are the most important factors in shorty headers. The larger the engine and the higher its RPM, the larger the primary tubes should be.

Some headers for inline four and V8 engines will have the primary tubes go into a pair of secondary tubes before coming together into a collector. These are known as Tri-Y headers due to their shape. This design fits well into a cramped engine compartment and can produce excellent low RPM torque. Unfortunately, they do not perform as well at high RPM as conventional four-into-one header designs.

Headers often will glow red hot at full throttle. This heat can make headers rust very quickly unless you take precautions. Plain steel headers should, at the very least, be painted with a high-temperature paint designed especially for exhaust systems. Some headers are shipped with a paint that is just to prevent them from rusting in transit and will need to be repainted after this paint comes off. If you want to use painted headers, any scratch should be touched up promptly. Ceramic coatings are an even better choice—these resist rust and heat, and also are hard enough not to be easily scratched. Cheaper ceramic coatings aren't entirely rust proof, however, particularly if the manufacturer didn't coat the inside.

Some headers are made from stainless steel. Stainless steel is not only corrosion resistant, but stronger at high temperatures than ordinary steel tubing. Unlike ceramic coatings, however, stainless steel will take on a bluish color from the heat. Note that when a stainless steel header reaches operating temperature for the first time, any fingerprints or patches of oil on them will burn into the finish. To keep these from marring the underhood appearance of a show car, clean the headers with rubbing alcohol before installing them, and then give them another cleaning before you first start the engine. Some grades of stainless can still rust—just not nearly as fast as normal steel.

If you are shopping for a set of headers for an older car, you may see fenderwell headers for sale. Installing these headers will require removing or cutting the inner fender, but will leave more room free in the engine compartment. This is not much of a problem on a car with a full frame, which is separate from the body. On cars like this, the inner fenders are little more than splash shields to keep road grime off your engine. On cars with unibody construction, however, there is no separate frame and the car body acts as the frame. The inner fenders in such a car are part of the car's structure. Cutting holes in the inner fenders is not only a change that is difficult to undue, but it weakens the structure of the car and may hurt its handling. If you want to be on the safe side, do not install fenderwell headers in any car where the inner fenders cannot be unbolted. If a manufacturer offers both headers that require cutting the inner fenders and headers that do not, they will frequently call the latter sort underchassis headers.

Fuel System

The fuel takes its own path from the fuel tank to the engine. Most fuel-injected engines have an electric fuel pump mounted inside the gas tank itself. This pump sends fuel through the fuel lines to a filter. After leaving the filter and traveling through more fuel lines, the fuel arrives at a fuel rail, a large tube mounted on the intake manifold. The fuel rail connects to the tops of the fuel injectors, a set of electronic valves that open or close to let a precise amount of fuel into the intake manifold.

At the other end of the fuel rail is a fuel pressure regulator. This contains a valve that opens and closes to keep a steady pressure level in the fuel rail. The valve stays closed below the required fuel pressure. Once the fuel reaches the required pressure, the valve opens. This lets pressure drop by sending excess fuel down a return line back to the fuel tank. Some newer fuel systems use a returnless fuel rail, which requires a different sort of regulator that sits upstream of the fuel rail or an electronically controlled pump that maintains a set pressure. Older engines may use throttle body injection, which uses one or more injectors in an assembly that also includes the throttle. Other variations put the fuel pump outside the fuel tank or combine a low-pressure pump in the tank with a high-pressure pump attached somewhere under the car.

Throttle body injection is close to a self-contained EFI fuel system and a simple way to replace a carburetor. (*Photo courtesy Edelbrock.*)

An engine control unit. (*Photo courtesy DIYAutoTune.com.*)

A mechanical fuel pump. (*Photo courtesy Edelbrock.*)

The fuel injectors need a control system to tell them when to open and close. While a few engines have used mechanical systems, most fuel injection systems control the injectors with a computer known as the engine control unit or ECU. This computer monitors input from a network of sensors that measure engine temperature, the engine RPM, and the amount of air flowing into the engine. The ECU uses this information to determine when to open and close the injectors so as to deliver the right amount of fuel.

Carbureted engines can use a simpler fuel system, or at least one that seems simpler until you try taking apart the carburetor and figuring out how it works. The most common design draws fuel out of the tank with a simple pickup tube, using a mechanical fuel pump mounted on the engine. The fuel pump has its own sort of pressure regulator built into it, requiring no return lines. After leaving the pump, the gasoline travels through a filter and arrives at the carburetor. The carburetor combines the fuel with the incoming air.

Supplying More Fuel

Since making more power often means burning more fuel, sometimes the fuel system needs to be upgraded to deliver additional fuel. The usual parts to consider are the fuel pump and, if you have an injected engine, the injectors. The other parts of the fuel system sometimes need upgrading too, either for performance or for safety.

HOW TO SIZE FUEL INJECTORS AND FUEL PUMPS

The parts of the fuel system must be sized so that they can supply the engine with enough fuel. Injectors or fuel pumps that are too small will make the engine run too lean at full throttle, which can cause serious damage. On the other hand, injectors have a minimum amount of fuel they can supply, and ones that are too large will cause the engine to run too rich at idle. The amount of fuel you need can be estimated from the amount of horsepower you hope to make, as well as whether you plan to use forced induction. Fuel system components may be sized in gallons per hour (gph), pounds per hour (lb/hr), or cubic centimeters per minute (cc/min).

To get a basic estimate of how much fuel you need, multiply the expected horsepower by 0.45 to get pounds per hour, 0.075 to get gallons per hour, or 4.73 to get cc per minute. With a turbocharged or supercharged engine, add 22 percent more fuel to the base amount. This is the amount of fuel the engine actually needs. Since you don't want a small drop in voltage or other problems to give the engine any less fuel, multiply this by a safety factor of 1.3 to 2, depending on how safe you want to play it.

To size the injectors, divide the amount of fuel flow the engine needs by the number of fuel injectors. Next, this number needs to be adjusted to allow you to let the injectors run only 80 percent of the time, leaving the rest of the time for them to cool off. To do this, divide the number by 0.8.

For example, suppose you are looking to get 300 hp out of your turbocharged Eclipse. If this engine were naturally aspirated, it would need a fuel flow rate of at least 135 lb/hr. Since it is turbocharged, it will need 22 percent more, for a real fuel flow of 165 lb/hr, and you'll want a pump that can move at least 215 lb/hr, probably more (the wiring on turbo Eclipses is notorious for not supplying the pump with enough power—you'll also want to go with a heavier-gauge wire). This engine has four injectors, so dividing 165 by 4 and then by 0.8 tells us that this Diamond Star will need injectors rated for at least 52 lb/hr. This is the same as the popular 550 cc/min metric-sized injectors. The original injectors are rated for 450 cc/min—not quite enough flow for this level of power.

Changing the fuel pressure can change the flow rate of your injectors. The size of the change can be estimated with this formula:

$$\text{New injector flow rate} = \text{Old injector flow rate} \times \sqrt{\text{New fuel pressure} / \text{Old fuel pressure}}$$

For example, if the injectors are rated to flow 450 cc/min at 43.5 psi and we turn the fuel pressure up to 65 psi, this can squeeze 550 cc/min out of the injectors. So the Eclipse can keep its original injectors after all if we install an adjustable fuel pressure regulator and the injectors can handle the extra pressure. The point at which injectors no longer work well can range from 60 to 75 psi, depending on the design of the injectors.

Fuel pumps come in two basic sorts: mechanical and electric. A mechanical fuel pump is compact, lightweight, and often comes with a built-in fuel pressure regulator. Most mechanical fuel pumps do not work with fuel injection.

Electric fuel pumps deliver a continuous flow of fuel that is suitable for fuel injection, although not all electric fuel pumps can work with electronic fuel injection (EFI). Usually, pumps rated for 20 psi or less are intended to be used with a carburetor, while fuel pumps meant for fuel injection put out 45 psi or more. A few systems use throttle body injection, where the injectors are mounted at the throttle body and run at lower pressures, around 18 to 25 psi. Electric fuel pumps virtually always need an external fuel pressure regulator, except for a few models with built-in regulators.

Many modern cars have the fuel pump located in the gas tank. This keeps things quiet and makes sure the fuel doesn't have to flow uphill to the fuel pump, although it can make replacing the pump a hassle. Many aftermarket pumps are "universal" designs meant to be bolted somewhere to the car body instead of in the gas tank. However, you can buy high-capacity in-tank fuel pumps from companies like Walbro. Some budget-minded hot-rodders have been known to use external fuel pumps from cars like the Nissan 300ZX or late '80s Ford pickup trucks instead of buying high-dollar aftermarket items, and run two pumps in parallel if they need more flow than one pump can provide.

External fuel pressure regulators have several features to check for if you need a new one. Obviously, you will want one that works in the pressure range your engine needs, usually around 5 to 9 psi for a carburetor and 40 psi or more for fuel injection. You'll also want to make sure its maximum flow rate is more than the fuel your fuel pump can flow—if the regulator is too small, the fuel pressure will rise, particularly at idle. After that, the most important is adjustability, which can be used for tuning or to make a fuel-injected engine behave like it has slightly larger injectors. Check to see whether the regulator requires a return line running back to the fuel tank or not. A normal regulator on an injected engine will add 1 psi of fuel pressure for every psi of boost pressure so that the air pressure does not interfere with fuel delivery. If you are running a carburetor with a turbocharger or supercharger and using what is known as a blow-through setup, you will need to connect a hose to send boost pressure to a pressure port on the regulator.

One expensive type of regulator is known as a rising-rate fuel pressure regulator. This one adds more fuel pressure as boost pressure increases, so it might add 2 psi (or more) of fuel for 1 psi of boost. This can be a cheap way of tuning an EFI system to add more fuel under boost, but is not as accurate as retuning the computer.

If the original fuel injectors cannot supply enough fuel, the best solution is to replace them with larger ones. Larger injectors can be bought from the aftermarket, but in many cases it is possible to find larger fuel injectors from another car. There are two important considerations when looking for larger injectors besides their flow rate. One is the shape of the injector; the injector must physically fit your engine. Since most injectors are made by a few companies (Bosch, Lucas, and Nippondenso being the biggest players), many times injectors from one brand of car will swap into another. You will need to make sure before ordering. The second consideration is an electrical property known as impedance. Installing injectors with a lower impedance than what your ECU is designed to handle can burn out the ECU. In most cases, installing larger injectors will require reprogramming or replacing the original ECU.

There are a few alternatives to installing larger injectors, particularly if you need just a little more fuel. One is to put in an adjustable fuel pressure regulator. Turning up the pressure will cram more fuel through your injectors. As a general rule, do not use less than 28 psi of fuel pressure, or more than 60 to 75 psi. The exact amount you can get away with will depend on your injectors. Naturally, this rule does not apply to the low-pressure injectors sometimes used with throttle body injection.

Aftermarket fuel rails may look impressive, but few modified engines really need them unless it's to fit more readily available injectors. In a few cases, a stock fuel rail may not be able to deliver enough fuel for a modified engine, but usually the first areas in need of upgrading are likely to be the injectors and fuel pump. Well-informed car clubs and tuners are likely to know if a particular engine has fuel rail issues, as well as possible remedies.

Fuel filters seldom need to be upgraded for performance, although if you are converting from carburetors to fuel injection, you may need a new filter to cope with the different fuel pressure. Safety is a key concern when selecting a fuel filter. Fuel filters have several case styles. The safest sort is metal. A fuel filter in a transparent glass case may look pretty, but is also likely to break open in an accident. Safety inspectors officially view these sorts of filters as fires waiting to happen. Most racing events will not allow any car on the track with a glass fuel filter.

Beyond safety, you'll need to make sure the fuel filter can handle the fuel flow you need, and that it can catch small enough debris. The smallest debris a fuel filter will catch is measured in microns. Carburetors need a filter that can keep out 40 microns or smaller, while fuel-injected engines need even better filtering to catch particles as small as 10 microns.

Fuel hoses and fuel lines are also an important safety consideration. While metal fuel lines can usually take almost any pressure a normal fuel pump will put out, rubber fuel hoses are rated for a maximum pressure. Hoses designed for carbureted fuel systems are not strong enough to take the high pressure of most fuel injection systems. Rubber fuel hoses can burst or wear through from rubbing against metal parts. Consequently, many racing safety inspectors allow only two feet of rubber fuel hose or less on a car (and, of course, no amount of low pressure fuel hose is safe to run on a high-pressure fuel injection system, whether the inspector sees it or not). If a car has too much rubber hose, it can be replaced with solid metal fuel line or braided metal hose.

There is no limit on how much braided metal hose a car can run if you use an approved brand of hose, but these will require special end fittings, available from the hose supplier. If you have added a very large amount of power to your car, it may be a good idea to install larger diameter fuel lines.

Some serious racing events may not allow using the original fuel tank. Instead, the regulations require a fuel cell, a special sort of fuel tank that contains a sponge-filled rubber bladder. Even if a fuel cell is smashed open, the fuel can only leak out of one slowly. Racing fuel cells do not always have provisions for a fuel gauge, and are usually one-size-fits-all boxes meant to be installed in the trunk. A trunk-mounted fuel cell is not the most practical choice for street use. Some companies do make custom fuel cells that install in the same location as the original tank. This can be practical on the street, but is rather expensive.

CHAPTER 5

Engine—Internal Work

Getting Inside the Motor for Bigger Power Gains

In the previous chapter, we looked at how to get fuel and air into and out of the engine. Now, we'll look at how the engine uses these ingredients to make power. Nearly all gasoline-burning cars today are powered by four-stroke piston engines. "Four stroke" means that the engine makes power by carrying out four different processes, known as *strokes*. Piston engines use pistons that move up and down in cylinders to carry them out, similar to the mechanism inside an old-fashioned bicycle pump. The pistons are attached by a set of connecting rods to a crankshaft at the bottom of the cylinder that controls their motion and transmits power from the pistons to where it can do useful work.

The first stroke is the intake stroke, where the piston moves downward to suck in a mixture of air and gasoline. This comes in through a valve (or sometimes more than one valve) in the cylinder head, the assembly at the top of the cylinder. The intake stroke is followed by the compression stroke, where the valve closes and the piston moves upward, compressing the air and forcing it into an area of the cylinder head known as the combustion chamber. A spark plug fires to ignite the fuel, starting the power stroke, where the resulting explosion forces the piston down. Once the piston reaches the bottom of the cylinder, another valve opens in the cylinder head, and the piston moves up to force the air out the exhaust. The last stroke is, appropriately enough, called the exhaust stroke.

There are three ways an engine can make more power: Burn more fuel, burn the fuel more efficiently, or find things the engine does that require power and reduce how much power is needed. There is little that can be done on a modern engine to make it burn fuel more efficiently that the factory has not already done. The engineers at the factory already went to great lengths to find out how to make the engine get as much energy as reasonably practical from every drop of gasoline, and outthinking them will not be easy unless some compromise for emissions rules forced them to leave some power on the table. Older engines can sometimes benefit from newer technology, however, as there are cases where you can add modern fuel injection and the latest high-efficiency cylinder head technology to old classics.

In this chapter, we'll take a closer look at what's inside the engine. (*Photo courtesy Edelbrock.*)

Most performance mods seek to allow the engine to burn more fuel by letting it draw in more air, since the amount of fuel an engine can burn is limited by how much air it can pull in. How much air the engine can pull in is determined by the size of the engine, the speed at which it can turn, and the amount of restriction placed in the path of the air. All of these factors can be changed with the right parts. The engine size is also

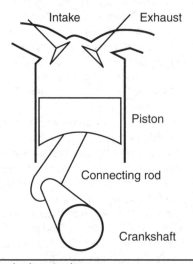

Diagram of a piston, intake, and exhaust valves.

known as its displacement, and is determined by the number of cylinders, the diameter of the pistons, and the stroke length, which is how far the piston moves during each stroke. Displacement is typically measured in either liters or cubic inches.

DISPLACEMENT AND VOLUMETRIC EFFICIENCY

Using the term "displacement" is a carryover from pump design, where the displacement of a pump is the amount of water or air that the pump can move from one place to another in one cycle. The displacement of an engine can be calculated with the following formula:

$$Displacement = (Bore/2)^2 \times p \times Stroke \times Number\ of\ cylinders$$

Using inches in this formula will give the engine size in cubic inches, while using centimeters and dividing the result by 1,000 will give the engine size in liters. One liter is equal to 61 cubic inches.

On a four-stroke engine, 1 cycle is 2 turns of the crankshaft. In theory, a 2-liter engine would draw in 1 liter of air every time you turned the crankshaft. In reality, it might only pull in 0.8 liters. The ratio of how much air an engine actually pulls in to the amount its size indicates it should pull in is known as *volumetric efficiency*, or VE. The 2-liter engine in this example has a VE of 80 percent. The volumetric efficiency changes with RPM. It is usually less than 100 percent, but can rise above this mark for some well-designed racing engines in a narrow RPM range, as well as for engines running a turbo or supercharger.

The rotary engine in this RX-7 still uses a four-stroke cycle, but it uses spinning triangular rotors in place of pistons.

There are a few oddities in the engine world. Mazda brought out the first rotary-engined car sold in the United States with the RX-2, and continued that design with a series of "RX" cars that ended with the RX-8. The rotary engine, also known as a Wankel engine, replaces the pistons and cylinders with a set of spinning triangles in a housing defined by a complicated mathematical shape. Rotary engines respond in a similar way to intake and exhaust work as piston engines, except they behave as if they have twice the displacement. Internal work is a different story, as virtually nothing inside the engine resembles anything found on a piston engine. Some European cars use two-stroke engines designed to fire each cylinder once per crankshaft revolution. Depending on the design of a two-stroke engine, it may not respond to mods in the same way as a four-stroke.

Cylinder Heads and Camshafts

The air flows from the intake manifold into the intake ports of the cylinder head, which is the assembly at the top part of the cylinder. The cylinder head has an arrangement of valves that open to allow the air from the ports into the cylinders. Normal engines use poppet valves here, which are metal circles attached to long stems. The circle fits into a circular area of the cylinder port and moves downward on its stem to open and let the air into the cylinder. These valves are held closed by valve springs, and pushed open by a device called a camshaft. Since the camshaft must spin in sync with the pistons to make sure the valves open and close at the right time, most engines use a belt, chain, or system of gears to turn it.

There are many different places to put a camshaft, and several ways to use it to open the valves. In all cases, the camshaft is a rotating metal cylinder running the length of the engine with a set of lobes on it, spinning half as fast as the crankshaft. Components known as *lifters* or *followers* stay in contact with the lobes, and move up and down as the lobes push on them. The shape of the lobes determines when the valves will open and close.

The oldest gasoline engines used a camshaft located in the engine block to operate valves located in the block itself, which often doubled as an intake manifold. In many cases, the lifters were simply cylindrical pieces of metal sitting in round bores that pushed directly on the valves. The cylinder head was often little more than a flat piece

A V8 cylinder head. The valves and their valve springs are visible at the top, but the rocker arms have been removed. (*Photo courtesy Edelbrock.*)

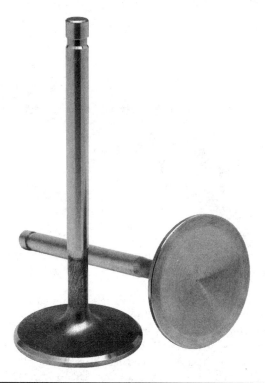

Poppet valves. (*Photo courtesy Comp Performance Group.*)

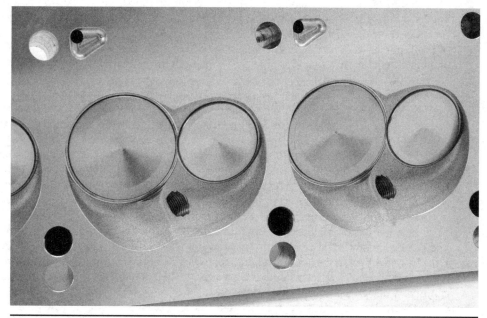

Close-up of the valves in a cylinder head. (*Photo courtesy Edelbrock.*)

A camshaft. (*Photo courtesy Comp Performance Group.*)

of metal with water passages inside it, so these engines are known as flathead engines. While a flathead engine is simple and compact, this design does not allow for very efficient combustion chamber designs.

The next step was to move the camshaft, lifters, and valves into the cylinder head, creating what is known as a single overhead cam (SOHC) engine. Some designers opted to add a second camshaft, making a dual overhead cam (DOHC) engine. Shortly after these designs, a more complicated design appeared, using a single cam in the block to open valves in the cylinder head through an arrangement of pushrods and rocker arms. This design is known as a pushrod engine. While flathead engines virtually always have one intake and one exhaust valve per cylinder for a total of two valves per cylinder, the other three designs make it possible to use two intake valves or two exhaust valves per cylinder, or sometimes more. Often, two small valves can create less restriction than one big valve. Engines that combine pushrods with more than two valves per cylinder are rare, but they do exist.

Today, the only new flathead engines appear in lawn mowers and other very small, very cheap engines. Pushrod, SOHC, and DOHC engines each survived for their own reasons. Pushrods allow for smaller cylinder heads than overhead cams, meaning that a large engine can be squeezed into a small package. A massive 7-liter pushrod V8 can often be as small on the outside as a 4-liter DOHC V8. Unfortunately, accommodating the pushrods in the cylinder head often means the designer must make compromises that can restrict airflow. The weight of the pushrods and rocker arms also can cause problems if the engine is spun at high RPM. The relatively small intake and exhaust ports work quite well to make torque at low RPM.

Single overhead cams do not have the same RPM limitations as a pushrod engine, and this design gives the designer more freedom to design a well-flowing cylinder head. However, needing to work both the intake and exhaust valves off one cam still creates compromises. Using two camshafts and four valves (two intake, two exhaust) per cylinder can give a designer the most leeway to design a high-performance cylinder head. The only downsides are that the extra parts add more cost and weight, and that a cylinder head designed for good airflow at high RPM may not perform as well at low RPM.

An old-fashioned flathead V8; note the cylinder heads are much smaller than on a modern engine.

When considering how well a cylinder head might perform, keep in mind that the theoretical advantages only become real advantages if the designer correctly uses them. There are pushrod engines that rev surprisingly well and overhead cam engines that have very restrictive cylinder heads.

Aftermarket Camshafts and Valvetrains

Camshafts are crucial in an all-out naturally aspirated buildup. The choice of camshaft is closely tied to the RPM where you want maximum power; in fact, many cam manufacturers describe their camshafts in terms of the RPM range where the cam works best. Most performance cams will make the engine breathe better at high RPM at the expense of performance at low RPM. Therefore, an engine builder needs to pick a camshaft with care. A cam that is too "big" for street use is certain to hurt acceleration at typical cruising RPM and cause rough idling, while the high RPM gains the camshaft can make will not happen without the appropriate other mods. Camshafts are typically measured in terms of duration and lift, with overlap and lobe separation angle also being listed on their specs.

The duration of a cam refers to the distance the crankshaft must turn, in degrees, between the time the valve begins to open and the time the valve closes completely. There are two kinds of duration numbers that you will often see: advertised duration, and duration at 0.050-inch valve lift. Advertised duration is measured from the time the valve starts moving to the time it stops, but manufacturers often have slightly different ways of measuring this. To standardize measurements, many companies also measure the duration from when the valve opens by 0.050 inch until the time the valve is all but 0.050 inch closed.

An engine that opened its intake valve at the very beginning of the intake stroke and closed its intake valve at the very end of that stroke would have a 180-degree intake duration. In ordinary engines, the valve is opened before the beginning of the stroke and closes after the end of the stroke. This sounds like it would push exhaust out the intake valve, drive the air from the intake stroke out of the cylinder during the compression stroke, and have similar troublesome effects. If you pick a cam with too long a duration, this can indeed happen at low RPM. However, very little air can actually move through the valve when it is just barely open. This means that at high enough RPM (where there is little time for the air to move through the partially open valve), these effects will be minimal. The advantage of a long duration, however, is that it allows you to run significantly more valve lift.

Lift is the furthest the camshaft can open the valve. Camshafts with more lift typically require more duration because there are limits on how fast the valvetrain can open and close valves. Opening or closing a valve too quickly will require more force and put more stress on the valvetrain. Most companies measure the lift at the valve, but some measure how high the lobe rises above the cam's base circle. On an engine that lacks rocker arms, these two measurements are the same.

In some cases, two camshafts with the same lift and duration will still allow a different amount of air into the cylinder. This is because of what is referred to as the area under the curve. Some camshafts may open the valves faster and hold them close to the peak lift for a longer time, allowing for more airflow without the consequences of too much duration.

The two measurements of lobe center angle, or LCA (sometimes known as *lobe separation angle* or *lobe displacement angle*), and overlap are closely related. Overlap is the distance the crank rotates, in degrees, while the exhaust valve has opened but before the intake valve can close. During this time, the momentum of the exhaust gasses leaving the cylinder can help pull air in through the intake valve. This can go too far in the case of an engine with large amounts of overlap running at low RPM, and suck the incoming air right out the exhaust valve. This effect is usually to blame for cases when exhaust mods reduce an engine's low RPM performance. LCA is the angle between the centers of the intake and exhaust lobes, and the smaller the LCA, the greater the overlap angle. More overlap can improve power at high speed at a cost of low RPM torque, while a cam with the same lift and duration but less overlap will trade maximum horsepower for better drivability. The low-overlap cam is also less likely to have a high-performance exhaust affect low-end torque.

The best aftermarket cams are made from brand-new cam cores, the metal casting that is shaped into a finished camshaft. In some cases, however, no new cores are available for an engine. In this case, a cam company can regrind a new lobe shape onto the lobes of a used cam. The cam grinder can either grind the base circle down further, or weld new metal onto the lobes and grind down the new metal. In either case, a reground cam will benefit from hardening through chemical or heat treatment.

In some cases, the effects of a camshaft can be altered by advancing it so the valves both open and close sooner, or retarding it so the valves open and close later. To a certain extent, advancing the cam will move the best performance downward in the RPM range, while retarding the cam will move it upward. This can allow someone who has installed a cam that is too "big" to compensate by advancing it, or let an owner looking to move the revs up try retarding the cam. The effects of this are limited, however: after adjusting the cam too far in either way, performance will suffer throughout the rev range.

On most pushrod engines, the only way to adjust the cam advance is with a complicated system of changes to the timing chain sprockets. Overhead cam engines, by contrast, often can use an adjustable cam gear that allows a tuner to advance or retard the cam with just a few screw adjustments when the appropriate covers are removed. Dialing in an adjustable cam gear requires dyno tuning.

Installing a camshaft is very involved and requires taking apart a sizable amount of the engine. On overhead cam engines, you can often lift the cam out of the cylinder head after removing the timing belt, valve cover, and the caps that hold the cam in place. Pushrod engines, however, require pulling the camshaft out of the front of the block. Since the camshaft is nearly as long as the engine block itself, this requires either partially disassembling the front end or removing the engine to install one. On front-wheel-drive cars with pushrod engines, pulling the motor is usually the only option. If you have the engine out of the car, you can (and should, if at all possible) use a special tool called a degree wheel to make sure the camshaft has the correct amount of advance.

Some recent engines feature different forms of variable valve timing. Honda's VTEC is the most widely known. It uses two separate camshaft lobes with different lift and duration, and employs a hydraulic system that switches between the two sets of lobes at a predetermined RPM point. Other systems use what is essentially a computer-controlled version of adjustable cam gears to advance and retard the camshaft. Some newer systems incorporate both. Either one can be tweaked by reprogramming the ECU or installing an aftermarket control system.

One important warning when working with camshafts or other valve gear is that increasing valve lift can reduce the clearance between the valve and the piston. Some engines, known as interference engines, already have so little clearance that if the timing belt (or timing chain) breaks, the pistons will crash into the valves, with catastrophic results. If you have increased your valve lift, be sure to measure the new piston-to-valve clearance. If the valves will now hit the piston when the piston is at the top of its stroke and the valve is at its full lift, you will have to pay careful attention to changing your timing belt at regular intervals or making sure your timing chain does not wear out. If the interference is bad enough, you may need to cut notches known as valve reliefs in the piston to keep the valves from hitting the pistons during normal operation.

The camshaft and cam gears are not the only part of the valvetrain that can be replaced with aftermarket parts. The cam will generally not push against the valves directly, but instead typically pushes against parts called lifters. Some engines may use other terms for these parts, such as buckets, lash adjusters, tappets, or followers. In some SOHC engines, the cam acts against a set of rocker arms.

Lifters come in several types, and it is important to use a cam with lifters that are designed to work with it. One basic difference in lifter design is that some use a roller, while others have the cam pushing against a flat surface covered with oil. Roller lifters often allow for more lift and can be reused when installing a new cam, while flat lifters must be replaced any time you replace a cam. Another difference is that some lifters, known as *hydraulic lifters*, use oil pressure to adjust for parts expanding as the engine warms up. Solid lifters lack this ability and require setting a small gap at some

Hydraulic flat tappet lifters. (*Photo courtesy Edelbrock*.)

The link between them keeps the rollers on these roller lifters pointed in the right direction. (*Photo courtesy Comp Performance Group*.)

place in the valve system to compensate. This means occasional adjustments, and can make an engine noisier, but allows for faster opening and closing rates. These types can be combined, resulting in descriptions like hydraulic roller lifters or solid flat tappets.

The camshaft pushes the valves open, but the valve springs push the valves back shut. Aftermarket cams may require new springs if they either push the valves so far open that the stock springs would bind (have their coils crash together) or the stock springs cannot close the valves quickly enough. In either case, installing a cam with the wrong springs can damage the engine. To close the valve faster will either require a stiffer spring or a lighter one. Most of the time the aftermarket springs are simply stiffer. The downside to stiffer springs is that they make for more power-robbing friction and more wear, so installing valve springs that are too stiff for your cam can almost be as bad as installing ones that are not stiff enough. Follow the cam manufacturer's recommendations when it comes to valve springs.

Rocker arms come in several sorts. One important difference is adjustable versus nonadjustable. Adjustable rocker arms can work with solid lifters, while nonadjustable rocker arms must be used with hydraulic lifters only. Some rocker arms are mounted on studs, while others are mounted on a steel shaft. The stud leaves the arm free to pivot in several directions, so they are used in combination with pushrod guide plates to prevent the rocker arms from getting turned sideways. These plates restrict the motion of the pushrods, and the pushrods restrict the motion of the rocker arms. Shaft-mounted rocker arms can take more RPM and produce less friction if they have been properly designed. Converting a cylinder head from stud-mounted rocker arms to shaft-mounted ones often requires machine work.

Sometimes the aftermarket offers high-performance rocker arms for pushrod engines. These can have several performance benefits, depending on what features they offer. Since a rocker arm is a lever, it can open the valve further than the actual cam lift. The ratio of valve opening to cam lift is known as the rocker ratio. A 1.5:1 rocker arm, for example, opens the valve 1.5 times as far as the camshaft moves the lifter. Some have different lengths to allow for more valve lift. A stiffer rocker arm body can have the

A more aggressive cam can call for new valve springs. (*Photo courtesy Comp Performance Group.*)

Shaft-mounted rocker arms. (*Photo courtesy Comp Performance Group.*)

same effect because a rocker arm can bend slightly, reducing how far it pushes the valve open. Roller bearings and roller tips can reduce friction and wear.

One last feature to look for on rocker arms is lighter weight, or simply concentrating the weight toward the center. This allows more RPM and running softer valve springs. Note that if a company says their aluminum rocker arms are for racing use only, they're not just worried about emissions laws—lightweight aluminum rocker arms have a limited life span. While in theory a car using these can be driven on the street, race-only aluminum rocker arms must be checked for cracks often enough to make this an impractical idea.

The aftermarket offers many replacement timing chains for older pushrod engines. Most of these engines originally used what is known as an inverted tooth chain. Inverted tooth chains are quiet and offer low friction. Unfortunately, as an inverted tooth chain wears, the sprockets can wiggle back and forth. The effect on valve timing is a minor nuisance, but the effect this can have on a distributor often results in timing problems. Some original equipment timing chains make this problem worse by using nylon teeth on the sprockets that wear out rapidly.

Double roller timing chains reduce this problem by replacing the thin teeth with large rollers that surround the pins holding the chain together. While a new inverted tooth chain will perform just as well as a double roller timing chain, as the engine wears, the double roller timing chain will retain its performance better. A single roller timing chain is more economical but less strong. Beware of cheap timing chains that look like a double roller timing chain but place the pins directly on the sprockets instead of inside rollers. These chains have more friction and wear out rapidly. Insist on a true roller timing chain. The timing chain will come with a set of matching sprockets; you can't use inverted tooth chain sprockets with double roller chains, or vice versa.

The stiffest valve springs call for even tougher cam drives. A gear drive is stronger and more accurate than any sort of timing chain. The top-of-the-line gear drives use a

A double roller timing chain. (*Photo courtesy Comp Performance Group.*)

A gear drive. (*Photo courtesy Comp Performance Group.*)

series of three gears and often offer such features as an easily adjustable cam gear. The less expensive gear drives use a pair of free-floating gears between the cam gear and the crank gear. This requires less precision to build and cuts costs, but it allows a little bit of wiggle room between the gears. Timing gear sets have several disadvantages. They may require machine work to install, especially the sort with three gears. Gear drives are also extremely noisy. Not only will other drivers notice, but on cars with original equipment fuel injection, the sound they make can interfere with the computer, which often uses a microphone to detect knocking sounds.

Cylinder Head Mods

The cylinder heads frequently offer room for improvement, especially on older cars. On many older V8 engines and some European sports cars, it is possible to replace a cast iron cylinder head with a modern design. In some cases, enthusiasts looking for a new pair of cylinder heads for a '60s muscle car on a budget can find a set of better designed heads from a late-model engine in a junkyard. Some examples are the Chevy Vortec, Ford GT-40, or Chrysler Magnum heads, all of which were used on various '90s-era trucks and SUVs. Those with more money can spring for a set of aftermarket heads. These upgrade heads often offer superior combustion chamber designs for improved efficiency and new port designs for less restriction. Many are also made from aluminum for less weight and better heat transfer.

Even if you keep the original cylinder head, you can modify the cylinder head for more power. The most common sorts of cylinder head work are porting, installing larger valves, and milling. Porting means reshaping the intake and exhaust ports to reduce restriction. Milling refers to removing material from where the cylinder head joins to the engine block to increase compression. Milling a cylinder head or installing oversized valves requires the help of a machine shop.

Port work is one you can do at home. It typically involves a considerable amount of work using a die grinder or rotary tool. A good cylinder head porter will try to smooth out any sharp contours and keep the cross-sectional area as close to the same for the entire length of the port. Higher RPM use often demands removing more material. You can often buy templates that serve as a guide for most popular cylinder heads. If you can afford it, your engine is likely to benefit from having the porting work done by a shop that can test their work on a device called a flow bench to measure how much the restriction has been reduced.

A machine shop can cut the valve seats, the area where the valve sits when it is closed, to allow installing larger valves if they will fit. While working with the valves and valve seats, a machine shop can also perform some additional useful mods. One is known as a three-angle valve job, which smooths the areas around the valve seat to improve flow. A machine shop can also install several mods at this point to reduce wear. Older engines designed for leaded gasoline used softer valve seats that can wear out if they are run on unleaded, so a shop can install hardened inserts into the valve seats to prevent this. Bronze valve guides are another worthwhile upgrade if your cylinder heads do not have them already. Valve guides are simply tubes that the valves run in, and if they wear out, they can cause the engine to burn oil.

Milling the cylinder head is often a cheap way to increase the compression ratio, the amount that the engine compresses the air during the compression stroke. There are a few cautions when doing this. As with pistons and cam upgrades, it is important to check piston-to-valve clearance if you are going to mill the head. The thickness of the mounting surface is also important; there is a definite limit to how much can be removed. This number is different, depending on how thick the surface was to begin with. Last, milling the head moves some parts of the engine, and may need a few other changes to compensate. The most common problems are that some pushrod engines may need shorter pushrods, and V-type engines typically need to have the surface where the head bolts to the intake manifold also milled. Despite these extra changes, milling the cylinder head can still be a cost-effective way to increase compression.

WHAT, EXACTLY, IS COMPRESSION RATIO?

The compression ratio is one way to measure the amount an engine compresses the air. The compression ratio is a ratio of the volume of air moved by the piston to the volume of air above the piston when the piston is at the top of its travel. For example, if a 4-liter engine has eight cylinders, each cylinder moves 0.5 liters, or 500 cubic centimeters, of air. If there is a 50-cubic-centimeter space above the piston, you divide 500 by 50 to find that the compression ratio is 10.0:1. A higher compression ratio can make an engine run more efficiently, but it also increases the chance that the engine will have problems with detonation—where the fuel explodes before the spark plug can ignite it. Detonation can cause serious damage to an engine. It can be avoided to some extent by careful tuning and running a higher-octane fuel. If you try running 15.0:1 compression, you're going to have a hard time avoiding detonation unless you fill the tank with race gas, alcohol, propane, or something equally exotic.

A compression check measures the actual pressure in the cylinders when the engine is off. This measurement is known as cranking pressure. The reading on the gauge is not the same thing as the compression ratio at all. This pressure can be influenced by the camshaft design as well as wear and tear on the engine.

Sometimes engine builders will talk about cylinder pressure, which confusingly enough does not refer to either compression ratio or cranking pressure. Instead, this is the pressure in the cylinders with the engine running. Cylinder pressure is the best way to tell if an engine is likely to have problems with detonation. It is increased by adding more compression ratio, but other factors like an efficient intake design will also increase cylinder pressure.

Engine Blocks

After traveling into the cylinder head, the air enters the cylinders themselves. An engine may have any number of cylinders, and they can be set up in several arrangements. The most common designs are inline and V. Most automotive engines use four, six, or eight cylinders, but there are also three-, five-, ten-, and twelve-cylinder engines on the market. As their names imply, an inline engine has the cylinders all in a row, and a V engine has two lines of cylinders arranged in a V. Nearly all four-cylinder engines and some six-cylinder engines are inline, while most engines with six or eight are V designs. Other arrangements include flat or boxer engines, where the cylinders are arranged in horizontal pairs, and slant engines, which are simply tilted inline motors.

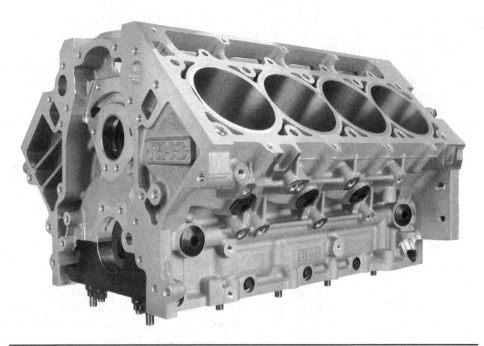

A V8 engine block. (*Photo courtesy Comp Performance Group.*)

The cylinders are typically part of a metal engine block, unless you are working on an air-cooled Porsche or VW, where each cylinder is a separate part. The pistons at the bottom are connected to a crankshaft using connecting rods. A connecting rod is a metal bar with a large hole at the bottom to attach to the crankshaft and a smaller hole at the top to attach the piston pin. The piston pin is a metal cylinder that holds the piston securely to the connecting rod.

The crankshaft and the big end of the connecting rod (yes, engine builders just call the end with the larger hole "the big end") are fitted with bearings. These are not the ball bearings you may be familiar with, but are simply strips of soft metal separated from the crankshaft by a thin film of oil. Since the oil keeps them away from the shaft, the bearings do not scrape against any metal when the engine is running correctly. The pistons, crankshaft, and connecting rods, along with the fasteners and bearings on them, are known as the reciprocating assembly or bottom end.

Upgrading Pistons, Rods, and Crankshafts

Much of the work with the reciprocating assembly requires a fully equipped machine shop. Installing new pistons the correct way often requires checking the bore, recutting the surface in a process called overboring, and polishing the cylinder walls with a device called a cylinder hone. There are cheap cylinder hones that attach to a drill, but it's rare to see a boring machine outside of a machine shop. Repairs to the crankshaft require similarly expensive tools. Many pistons also require a heavy-duty hydraulic press to attach them to connecting rods. In some cases, you may want to leave all the

Aftermarket forged pistons. The one on the left is for a diesel engine.

work to a machine shop. Other builders prefer to have a shop do the machine work and then finish assembling the engine themselves.

Although changes to the pistons can make extra power through more compression, the most important consideration when choosing aftermarket parts for the bottom end is durability. Aftermarket parts are often made stronger for high RPM or the additional stresses of forced induction. This can be very important if you are pushing the engine

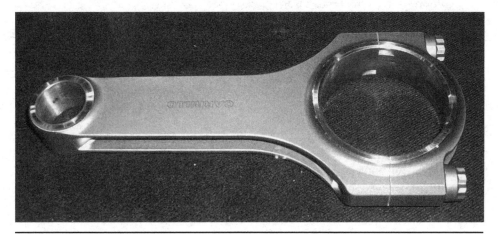

An aftermarket connecting rod.

significantly beyond the RPM range its designers intended, as an engine spinning twice as fast experiences four times the internal stress.

Pistons are usually offered in two main types: cast and forged. Both types are made from aluminum alloys. Cast pistons are inexpensive, often lightweight, and often have steel reinforcing parts cast in to limit how much they can expand from heat. The low expansion rate means that the pistons can seal more tightly in the cylinders, which reduces noise and can produce slightly more power. The disadvantage of cast pistons is their low strength, making them usually a poor choice for heavily modified engines or ones with forced induction: turbocharging, supercharging, or nitrous oxide. Cast pistons can live with forced induction as long as boost pressures (or the amount of nitrous) are kept low and the engine is well tuned to avoid detonation.

Not all cast pistons are created equal. Hypereutectic pistons are a type of cast piston made from a stronger alloy. Although not as strong as a forged piston, they are tougher than more conventional cast pistons. They do not work well with high-powered engines with forced induction, but for a moderately powerful, naturally aspirated engine (one without forced induction), they can work quite well.

High-performance engines frequently use forged pistons. Although more expensive, these offer the strength to survive high RPM or high-pressure forced induction. Forged pistons must run a large amount of clearance to deal with expanding when the engine warms up, and this can produce a considerable amount of noise and make the engine burn a little more oil until it reaches normal operating temperatures. Check the manufacturer's recommended piston clearance and go with a forging that can run tight clearances if the noise is an issue.

When selecting pistons, you may find you have a choice of compression ratios. Engines running forced induction will often run compression ratios as low as 8:1, while a high-revving, all-motor buildup may sometimes use a compression ratio as high as 12:1, or even higher on engines that burn race gas or alcohol. More compression will make an engine more efficient, but it will also increase the risk of detonation. If you plan to increase the compression ratio, use higher-octane gas. Very high compression ratios like 12:1 are only possible if the engine has a high-performance camshaft that keeps the valves open long enough that some of the compression "leaks out," or if the engine is running a fuel other than gasoline. Aluminum cylinder heads will also allow higher compression than their cast iron counterparts.

The cam profile affects the engine's compression, even though it does not change the compression ratio itself. A cam with a large amount of duration can cause the pressure to leak out on the intake stroke, as the valve is not completely closed yet. When running a very aggressive camshaft, an engine will need a higher compression ratio to compensate for this. If, for some reason, an engine builder detunes an engine by installing a much milder camshaft than stock, this can call for lowering the compression ratio.

A piston will not fit snugly enough in the bore to keep the air in the combustion chamber from leaking past the piston. Instead, a set of spring-loaded rings around the piston handle this job. Almost all modern engines use two compression rings at the top of the piston to hold back the air, and a wider ring below these two to keep oil from getting onto the top of the piston. These rings are usually made from cast iron. Some rings offer chrome or molybdenum ("moly") coating to better withstand high temperatures and wear longer. Chrome rings often require special machining and break-in techniques to get them to seal properly. Molybdenum rings do not have this problem, making them one of the most popular choices. Before installing new piston rings, it is necessary to hone the cylinder walls. This will ensure that the engine "breaks in" and the rings seal correctly to the cylinder walls. If there is any significant wear on the cylinder walls, you may need to have a machine shop bore the cylinders slightly larger and install oversized pistons meant for such repairs.

Connecting rods and crankshafts also come in both cast and forged varieties. In this case, the cast versions are usually iron, and most forgings are steel. Cast iron is cheaper, but forged steel is considerably stronger, and usually the choice for serious performance buildups. Unlike with pistons, forged rods and cranks seldom have any drawbacks besides cost. Some cars designed for performance or durability will use forged steel parts from the factory, but even then there may be stronger aftermarket replacements. Checking with experts on your engine can often turn up how much power your original parts can endure.

The bolts used to hold the connecting rods are quite important, as this area is under a tremendous stress. This is often the weakest point for original rods. Ordinary hardware store bolts just won't do; you will want bolts specifically made for connecting rods. If you are doing any work on the connecting rods, it pays to replace these with the strongest rod bolts you can find.

One option for those considering replacing the reciprocating assembly is a stroker kit. This uses a different crankshaft to lengthen the piston stroke, making the engine bigger. Such a kit may use a new aftermarket crank or one from a different, related motor. If no such crank is available, a machine shop can weld metal onto the rod journals (the points on the crankshaft where the connecting rods bolt on) and grind this metal into the shape of a new rod journal. Custom crank work isn't cheap. When adding a stroker crank, it is almost always necessary to change the pistons or connecting rods to get an acceptable compression ratio.

An engine comes with a few measures already taken to reduce vibration by designing the parts so that weight in one spot is balanced by weight elsewhere. The crankshaft uses a set of counterweights to bring vibrations down to an acceptable level. Some engines are internally balanced, having all the counterweights on the crankshaft itself. Other engines are externally balanced and require offset weights on the parts attached to the ends of the crankshaft. In some cases, a machine shop can take an externally balanced engine and balance it internally. If you have this done, be sure to replace the counterweighted parts on the outside with neutrally balanced ones.

Often, "balancing an engine" describes a more complicated process. To balance an engine, a machinist carefully removes weight from the right places on the pistons, rods, and crankshaft. The parts are carefully weighed to be sure that no piston weighs more than the others, that the rods all match up, and that there is no weight on the crankshaft that is not counterbalanced by weight in another spot. A balanced engine will run more smoothly and be more reliable at high RPM. This calls for a machine shop with appropriate balancing tools.

Blueprinting is a technique similar to engine balancing. The original design for an engine will rarely specify exact dimensions. The blueprints might call for a 0.25-inch hole to be anywhere from 0.247 to 0.253 inch in diameter, for example. This variation from the official dimension is called a tolerance. A blueprinted engine will have all of the parts machined to whatever end of the tolerances are most favorable to performance. This is usually done to engines built for racing in a class that places strict limits on modifications, but does not consider any part that meets factory specs to be a modified part. Like balancing, blueprinting is best left to experts.

The end of the crankshaft will also feature a device on the front pulley known as a harmonic balancer or damper. This balancer does not always compensate for off-center weight like other sorts of balancing (although it usually gets part of this job on an externally balanced engine). Instead, it absorbs the vibrations known as crankshaft harmonics. Each time a cylinder fires and pushes on the crankshaft, the crank will twist a small amount and then spring back. If left uncontrolled, the crank will keep twisting back and forth for several cycles, putting more wear and tear on itself. A harmonic balancer will absorb these twisting vibrations, making the engine run more smoothly and reducing the stress these vibrations put on the crankshaft. Aftermarket racing dampers are designed to cope with the additional stress of prolonged high RPM use.

Ignition System

There are many ways to send a spark to the cylinders. One of the oldest systems is known as breaker points. Breaker points are a mechanical switch that opens and closes as the engine turns. The points are mounted inside a distributor, a mechanical ignition control device. The distributor contains a mechanical system for determining when to trigger the points so as to fire the spark plug at the right time. When the points open and close, they generate a signal of 12 volt pulses.

The pulses run to a coil, which is often inside a housing about the size and shape of a soup can. The coil amplifies from 12 volts to 40,000 volts or more. This high voltage runs along a thick wire to a connection in the center of the distributor cap. The distributor contains a spinning metal rotor with one end connected to the center terminal. The other end connects to a circle of terminals at the edge of the distributor, one terminal for every cylinder. This sends a precisely timed high-voltage jolt to a wire leading from the distributor terminal to the spark plug, which ignites the air and fuel in the combustion chamber.

The time at which the spark fires is typically measured in degrees. The degrees are the angle of the crankshaft when the spark fires. They are typically compared to top dead center, the position where the crankshaft points straight down the middle of the cylinder and the piston is as far to the top as it will go. Timing is measured in degrees before top dead center (BTDC) or after top dead center (ATDC). The spark is not always fired at the same angle. Different conditions call for the distributor to advance the timing (make it fire sooner before top dead center) or retard it (by making it fire later, closer to top dead center). This timing requires great precision. At 6,000 RPM, having the timing off by *one millisecond* would mean firing the spark plugs 36 degrees from the correct ignition point.

While points work, a high-speed mechanical switch needs frequent maintenance. Many later designs improved on this system. The first improvement was to replace the points with a sensor inside the distributor, which was wired to an ignition module. This module can be as simple as a large switching transistor or as complicated as a capacitive discharge box. Some modules deliver a 12 volt signal to the coil, just like the points. Others, like the capacitive discharge, deliver higher voltages, around 400 volts.

INDUCTIVE DISCHARGE VS. CAPACITIVE DISCHARGE

Getting from 12 or 14 volts to the 40,000 or more volts needed to fire a spark plug is not easy. The key to this is the coil, which contains two lengths of wire wrapped around a core. The primary winding connects to the ignition controls, while the secondary winding connects to the high-voltage spark plug wires. Normally, the secondary winding wraps around the core over a hundred times more than the primary winding.

Breaker points and many stock ignitions use what is known as inductive discharge. In this case, the 12 volts flow through the primary winding of the coil and build up a magnetic field. This magnetic field stores energy. When the 12 volts turn off, the current stops and the magnetic field collapses. When the magnetic field collapses, the energy stored in it goes into creating a voltage spike. Since the secondary winding has the magnetic field running through hundreds of times more loops, the voltage spike is hundreds of times stronger in the secondary coil than the primary coil.

The key to an inductive discharge is to make the magnetic field collapse as quickly as possible, maximizing the voltage spike. The better the points are at opening fast or the amplifier is at switching quickly, the more volts the coil can deliver. Normal breaker ignitions use a small capacitor to prevent electricity from arcing across the points, since this arcing both makes the field collapse more slowly and puts wear on the points.

A capacitive discharge ignition is different. It generates a spark by storing energy in a capacitor rather than in a magnetic field. The circuitry charges the capacitor to around 400 volts and then sends the stored charge in the capacitor through the coil. The result is an intense but very short spark. A short spark may not have as much time to ignite the fuel, but it does allow firing multiple sparks per cylinder to make up for the short duration. A capacitive discharge ignition has a short recharge time, which is a big help if the engine has very little time to fire the spark. An all-out racing V8 turning at 8,000 RPM and using a single coil has less than two milliseconds to create each spark.

More sophisticated ignitions control the spark timing with the engine control unit (ECU) or a separate controller instead of with a mechanical system in the distributor. Some use a single coil and just keep the distributor as a switch to determine where to send the spark. Others remove the distributor entirely and use several sets of coils, triggered by separate ignition modules. There are various ways to arrange these coils, including attaching them directly to the spark plug to eliminate the spark plug wires. Using multiple coils not only removes the wear on the distributor; it also lets several coils charge at the same time, giving the coils more time to charge at high RPM. Many factory-original ignitions today are as hot as a serious racing ignition from the 1970s, and more accurate.

A Better Spark

Different cars respond differently to ignition mods. Lightly tweaked modern cars often do not benefit from any ignition upgrades other than using top-quality spark plugs and wires. A heavily modified engine may need changes to the computer to alter the timing settings, and adding forced induction can sometimes mean a car will need a more powerful ignition module. Older cars may benefit from having all or part of the features of newer ignition systems swapped over. Many owners consider the reduced maintenance alone enough to make removing breaker points worthwhile.

Recently, a bewildering variety of spark plugs have appeared on the market. However, most racing engines still run ordinary copper spark plugs. Some of the exotic new materials such as platinum have been known to improve plug life, but not usually performance. One plug tweak that can sometimes help is to increase the spark plug gap—the distance between the electrodes on the tip of the plug. This can often produce more horsepower if the ignition system has enough power to still fire the plug at the new gap. If the ignition is not powerful enough, this will cost horsepower and leave the tip of the spark plug covered with soot from incomplete burning. On the flip side, many engines that run forced induction may need a smaller spark plug gap to allow the spark to travel through the higher cylinder pressure.

Aftermarket spark plug wires often offer less resistance, which will deliver more current to the spark plugs, and more insulation, to keep the extra voltage from a high-powered coil from leaking out. Aftermarket wires often feature metal cores instead of the carbon cores in ordinary wires. Reports of testing aftermarket plug wires against the originals have shown mixed results—sometimes no gain, sometimes as much as 3 hp or so. In any case, spark plug wires wear out, and worn-out wires will cost you horsepower.

Aftermarket performance coils can increase the spark voltage. If looking for a coil, keep in mind that not all coils are designed to work with all ignition systems. A low-resistance coil will draw so much current if connected to an ignition system designed for a high-resistance coil that it may cause other parts of the ignition system to burn out. A full-tilt drag racing coil may deliver a very powerful spark, but only be designed to run for two or three minutes before turning off the engine and giving the coil time to cool off. Before buying a high-performance coil, make sure it is designed to work with your electronics. In some cases, you will want to buy a new coil when you buy an ignition module, and buy the coil and amplifier as a matched set.

On a car originally equipped with a distributor, a capacitive discharge ignition box can be a good upgrade if you are running it at much higher RPM than stock or are running forced induction. Many inductive discharge distributor setups lose spark energy at high RPM and may not be up to snuff if the RPM or cylinder pressure gets too high. Distributorless ignitions are better at keeping the same amount of energy at high RPM and less likely to benefit from a capacitive discharge ignition.

Cooling System

Unless you have an older Porsche, Corvair, or VW Beetle, you probably have a water-cooled engine. These engines use a water pump to circulate water through the engine block and cylinder head to carry away engine heat. The pump circulates hot water mixed with antifreeze to a radiator to release the heat, and then returns it to the engine. The engine does need to reach a certain temperature to run efficiently and keep its oil clean. A temperature-operated valve called a thermostat closes when the engine is too cold, allowing the engine to warm up.

The radiator requires a fan and a series of ducts or fan shrouds to channel air through it. The fan can be driven by either the engine or an electric motor. Some engine-driven fans use clutches to turn off the fan at either high RPM or high temperatures.

Cooling System Upgrades

A more powerful engine can often put out a bit more heat, and racing often puts more demand on the cooling system even if you haven't turned up the power very much. The aftermarket offers a variety of parts to help cool the engine better while wasting less power. Improvements in cooling can come from upgrading the radiator, fan, or even the liquid in the cooling system.

Although in theory a copper or brass radiator would provide the most cooling, many original equipment radiators that use these metals also use a lead solder to hold them together that impairs cooling. Aluminum radiators use welds instead of solder and often can provide more cooling for their size. Also, many aftermarket aluminum radiators are both larger for more cooling and lighter at the same time. Although there are custom-made aluminum radiators meant to fit specific cars, many of them are generic designs that can require a little drilling—and sometimes a little work with a hacksaw—to install. If you have an automatic transmission, be sure to use a radiator with provisions for a transmission cooler or heat exchanger. Since aluminum radiators have become increasingly common from the factory, a trip to the junkyard with a tape measure might turn up a useful upgrade. There have also been several companies offering inexpensive bolt-in radiators made in China appearing on the market lately. These may not be top quality, but ones from a reputable distributor will hold up fine for ordinary use.

Engine-driven cooling fans often use up a considerable amount of horsepower. Selecting the right fan can make a big difference in performance. One test actually found a poorly designed aftermarket fan on a V8 engine needed nearly 50 hp to turn it at full speed! Flex fans are designed to bend and theoretically use less power at high speed, but in practice many of them need more power than a stock fan. A thermal or RPM-activated clutch on a stock-type fan can often free up significant amounts of power. If using an aftermarket engine-driven fan, try to find one that fits in your original fan shroud, as running with no shroud makes a cooling fan less effective.

Often, an electric cooling fan will require the least amount of power. While there is no shortage of aftermarket fans, a trip to the junkyard may turn up something usable that can fit your radiator for less money.

There are several additives or alternate coolants that can transfer more heat. Two effective ones are Redline Water Wetter and Evans NPG Waterless Coolant. Changes to the cooling liquid can increase the boiling point so it does not boil at local hot spots or affect the surface tension to keep bubbles out of the cooling system.

Some cars do not have enough airflow through the radiator, and will benefit if you can get more airflow through it. This problem is especially common on cars with engines that are much larger than the original; a notorious example is a Ford Pinto with a 5.0 V8 stuffed under the hood. Improving airflow may require enlarging the grill openings; blocking off ways for air to flow into the engine compartment without going through the radiator; and adding vents, louvers, or ducts for hot underhood air to escape.

Oiling System

Many of the moving parts on an engine are only separated from other metal parts by a thin film of oil. With no oil, the parts would build up so much friction that they can weld themselves to each other. A normal street engine uses a relatively straightforward oil system. The oil collects in a pan at the bottom of the motor, below the crankshaft. A pickup tube in the oil pan draws the oil into a pump, which sends the oil through a filter and then through a series of passages to lubricate the engine.

Performance Oil Mods

You may be surprised to find that you can add power by changing the oiling system. Several horsepower are lost to friction in the bearings, and more are used up if too much oil sticks to the crankshaft. Switching to a thinner oil, particularly a synthetic one, can reduce friction. Further gains can come from installing a crankshaft scraper or windage tray in the oil pan, which removes the oil from the crank and returns it to the oil sump. Having the crank hit the oil can both churn the oil into foam and create more resistance to the crank turning.

In other cases, you may need to make a few upgrades to the oiling system to improve engine life under hard driving. One of the most important things is to make sure the oil in the oil pan does not slosh around during hard cornering, where it may slosh away from the pickup and keep the engine from getting any oil. A baffled or deep sump oil pan can help keep the oil in place and carry extra oil.

Usually, the engineers designed the stock oil pump to work fine even under racing conditions. A few very high-RPM engines can benefit from a high-volume oil pump. In most cases, though, the stock oil pump itself will be fine even for all-out racing.

A few street engines and many all-out road racing engines use a dry sump oiling system. This uses a larger, external, belt-driven oil pump to suck the oil out of the pan and into a separate oil tank, where a second stage in the pump sends the oil back to the engine under pressure. This both keeps the crank and rods from picking up too much oil and helps avoid oil starvation under hard cornering.

Wasting Less Power

The alternator, power steering pump, and other accessories require power to keep them turning. While it may not always be reasonable to remove these entirely, there are kits known as underdrive pulleys to turn them more slowly, using less power. These can pick up a small gain in horsepower.

One system that many enthusiasts would rather do without are the balance shafts. Usually used on four-cylinder engines and some V6 designs, these shafts spin a set of counterweights designed to cancel out the vibration of the crankshaft and pistons. Removing or disconnecting the balance shafts will free up a few horsepower, at a cost of making the engine feel a little bit rougher. V8 and inline six engines generally neither have nor need balance shafts.

Words of Caution

Internal engine work and other hardcore mods can turn your motor into a monster. Whether this is a monstrous amount of horsepower or simply a beast to drive in traffic will depend on how well you choose your parts. Parts that work poorly together can hurt horsepower, cause the car to run poorly, and shorten the life of your engine. Even a well-chosen set of naturally aspirated mods that makes an enormous improvement in horsepower may result in a horsepower gain only at the top of the RPM range and worse performance at low RPM. An all-out racing engine may not even be able to idle below 2,000 RPM. Building large amounts of naturally aspirated power has trade-offs, and getting greedy can make a car truly unpleasant as a daily driver.

One useful measure of how much you might be able to gain from internal work is the amount of horsepower per liter your engine already makes, also known as its specific output. The more you already have, the more difficult it will be to increase this number without resorting to forced induction. An American V8 from 1977, choked with the crude smog control equipment of its time, might make as little as 30 hp/liter. That same engine might be brought up to 75 hp/liter on a low budget with smog equipment removed and some good internal work. On the other hand, if you have a high-revving sport compact that already cranks out 100 hp/liter, more than doubling your horsepower without forced induction is not likely even on an all-out race motor. Getting even 30 percent more power out of this car and keeping it "all motor" is likely to give the car power that is not useful below 8,000 RPM, at the expense of power below 6,000. Adding more power could not only make the car feel slower, but in some racing events, it may actually be slower. In such cases, forced induction is likely to give more usable power.

If you want an all-out race motor, you will want to find a machine shop that specializes in building race engines, ideally one that works often with cars from your manufacturer. A shop whose bread and butter is repairing semitruck engines may not be familiar with the ins and outs of preparing a race motor.

Engine Swaps

Sometimes, the original engine just can't be modified enough to satisfy your thirst for power. The best option may be to transplant in a more potent engine, a popular mod since the days of swapping flathead V8s into Model T Fords. This choice requires a good deal of planning and research. It is also a wise idea to make this choice as soon as possible into a project to avoid spending good money on an engine you do not plan to keep. There are a few cases where you may be able to move mods from the existing engine to a new one, but this usually is not possible unless the engine is extremely similar to the original.

While an engine swap may be intimidating for a beginner, some swaps are harder than others. In some cases, you may only need to unbolt and disconnect the old engine and bolt a new one in. Other engines may require an aftermarket mount kit or swapping transmissions too. This may make for more time and money, but not necessarily a swap that demands more skill. Some swaps can present much more difficulty, from creating wiring difficulties to requiring modification to parts of the car's subframe. Uncommon swaps may require building unique parts just for your car. Even if you are planning to have a mechanic carry out the engine swap, knowing how much difficulty is involved can help you get an idea of what things may cost and how carefully you will need to search for a good mechanic if you're not doing the work yourself. Even a shop that routinely installs performance parts may balk at the thought of shoehorning a Lexus V8 into a Corolla.

Usually, the easiest swaps involve engines not only from the same manufacturer, but also from the same engine family. For example, the 1.6-liter Honda B16A can easily be replaced by a B18C from an Integra. A Chevy 307 V8 can be replaced with a 350. If your car model was offered with an engine in the same family as the one you want to install but you do not have that engine in yours, you may need to track down a few parts used to install the other engine option.

As a general rule, if you have a rear-wheel-drive American car from the '60s or later, it is not a good idea to try swapping in a V8 from a competing manufacturer. Not only is this likely to be more work and expense, but fans of your car brand are likely to take this as an insult. Earlier cars tend to be considered street rods, and often have somewhat more permissive unwritten rules.

A well-chosen engine swap can often get you not just more horsepower, but more available speed parts if you want to add more. For example, if you have a Mustang with a 2.8-liter V6, you will find considerably more mods available if you put in a 5.0 or its larger cousin, the 351 Windsor. However, an unusual engine and chassis combination may prevent you from installing some external mods. There are a lot of performance parts for a Honda B18C, but a header that fits if you have swapped one into your '78 Civic CVCC might not be one of them. This may require taking your car to a racing shop and having them build a custom header.

Many companies have sprung up to import used engines from Japan. A very expensive system of inspections in Japan often forces cars off the road early. Many cars there are scrapped long before Americans would consider their useful life over, so at one point, used engines from Japan were less likely to come from cars that had been driven into the ground. Another advantage of Japanese-market engines is that many Japanese manufacturers reserved some of their hottest engines for their own use. There are many interesting JDM (Japanese domestic market) engines not normally available in the United States, such as the Nissan RB25DET, a turbocharged inline six that is relatively easy to swap into most 1990s-era rear-wheel-drive Nissans. Some companies import entire "clips," which are basically the front half of the car cut off with a cutting torch. Buying a whole clip can give you many useful parts for swapping your JDM engine into a U.S.-market car, not to mention some useful customizing parts if the sheet metal is also in good shape.

If you are importing an engine from Japan, there are several issues to watch out for besides emissions inspectors. Although the engines often do have low mileage, Japanese cars are more likely to have been driven in heavy stop-and-go city traffic for short trips. This puts considerably more wear and tear on an engine than cruising for an hour at a steady 70 mph. The supply of many of the more desirable engines is not nearly as common as it was in the 1990s, and it's all too common for a "low mileage JDM engine" from a questionable supplier to arrive in need of a total rebuild. And just imagine how difficult it may be to find any replacement parts if you have imported an engine that does not share parts with a U.S.-spec relative.

When swapping engines, weight can be as important a consideration as fit. An engine may be heavier, or in some cases lighter, than the one it replaces. If you are putting a motor in a car built to handle, your suspension tuning must take this swap into account. Some swaps can produce a car so nose-heavy that it is not likely to be a good choice for anything other than drag racing. A Ford Pinto with a 7-liter V8 from a Lincoln may tear up the quarter mile, but a 2.3 Turbo swap out of a Thunderbird Turbo Coupe (which is a much updated version of the original Pinto motor) would be a far better choice if you want it to handle well. Plus, this is a much easier installation and gives you more room in the engine compartment.

Check your local emissions laws before performing an engine swap. Some areas may allow anything that passes a tailpipe emissions test. Other states may require that the engine meet all the specifications of a U.S.-market engine sold in the same year your car was built or later. A few areas even forbid installing an engine with more cylinders than were originally available for your particular car or installing a truck engine in a car.

One other important factor with an extreme engine swap is whether the rest of the drivetrain can handle the increased power and torque. Sometimes you may need a beefier transmission even if the engine bolts to the one you have. Axles and brakes may also need upgrades. Of course, you may need to upgrade these if you have done enough work on your original engine.

CHAPTER 6

Ultimate Horsepower

Turbocharging, Supercharging, and Nitrous Oxide

W hen working with a naturally aspirated, or "all motor," engine like the ones we talked about in the last chapter, the goal is to help the engine draw in as much air by itself as possible. For those not satisfied with the power an engine can make like that, the next logical step is to add mods that actively push even more air into the engine. The extra air allows the engine to burn more fuel and produce both more torque and more horsepower at every RPM. Since this approach forces air into the engine, it is known as forced induction. Some also call forced induction devices "power adders." Forced induction usually falls into three basic categories: turbocharging, supercharging, and nitrous oxide injection. Turbocharging and supercharging both use pumps to add extra air, while nitrous oxide injection supplies extra oxygen by adding chemicals from a tank of compressed gas.

Forced induction can make for incredible power gains, limited mostly by the strength of the engine, the octane of the fuel, and the amount of driveability needed. This method of building power is useful for those who want to add power without adding RPM to their engine. In some cases this may be because their engine has a long stroke and would not behave well at high RPM. In other cases, the engine is already close to "maxed out" and it is difficult to gain useful power while staying naturally aspirated. Or sometimes, you just might want more low RPM torque without an engine swap.

The most popular way to measure how much air a turbo or supercharger adds is to boost pressure, the level of pressure in the intake manifold. On a naturally aspirated engine, connecting a pressure gauge to the intake manifold would read zero, at best. Most of the time, you would have a negative reading, a partial vacuum. Pulling air through the filter, throttle, and other parts causes its pressure to drop below the pressure of the outside air. If you drilled a hole in the manifold, air would flow into the engine through that hole. That would be a vacuum leak, and these tend to give an engine difficulty adding the right amount of fuel.

This is not the case when you have a pump cramming air into the engine. In this case, the air pressure in the manifold goes into the positive range at full throttle. The reading

This Ray Barton Hemi makes well in excess of 1,500 hp using a pair of Precision turbos and a MegaSquirt-III control unit. It's not clear just how far above 1,500 hp it goes, since the car isn't able to get all the power down on a chassis dyno. (*Photo courtesy DIYAutoTune.com.*)

on the positive side of the pressure gauge is known as *boost*. If this engine had a hole in the manifold, air would escape out of the intake instead of flowing into the hole.

ABSOLUTE AND GAUGE PRESSURE

There are two common methods for measuring pressure: absolute and gauge. A device that measures absolute pressure compares the gas pressure with the pressure in a total vacuum, so a total vacuum would read 0 psi. Typical absolute atmospheric pressure is 14.7 psi, but it changes depending on altitude, temperature, and weather. Gauge pressure, on the other hand, compares the pressure being measured with the atmospheric pressure. Atmospheric pressure is always 0 on a scale that measures gauge pressure, no matter what the absolute pressure of the atmosphere is. Virtually all automotive gauges work with gauge pressure, as the name implies. Gauge pressure is also what you use to set fuel pressure. Many types of engine electronics, however, work based on absolute pressure.

Boost is not the only way to measure how much air a turbocharger or supercharger adds to the engine. Another measure is pressure ratio. To find pressure ratio, take the absolute pressure (not the boost pressure) in the intake manifold and divide it by the absolute pressure outside the manifold. For example, if the pressure is 28 psi inside the manifold and 14 psi outside, you would have a 2:1 pressure ratio. The boost pressure would be 14 psi.

While forced induction does put additional stress on the engine, a correctly tuned engine is usually under less stress than making the same amount of power with the same engine design and no forced induction. Although you might expect most of the stress on the engine to come from pressure in the cylinders, the greatest stress on internal engine parts actually occurs at the end of the exhaust stroke, if the engine is working correctly. At this point, the speeding piston must come to a dead stop and reverse direction. Since there is very little pressure in the cylinder to push down on the piston, the only force that can stop the piston comes from the connecting rod. This can pull the connecting rod with a force exceeding two tons, and the force increases with the square of the RPM. The result is that adding RPM is potentially more damaging to an engine than adding boost, if the engine is tuned correctly.

The problem is that the engine may not always be tuned correctly, particularly when you first start sorting things out. If an engine is not tuned correctly for forced induction or runs too low of an octane of fuel, it may have problems with detonation. Detonation is where the fuel burns unevenly and parts of it explode before the fire reaches them, sending shock waves through the combustion chamber. This can ruin head gaskets and pistons, or even damage bearings and connecting rods. Since tuning problems are more likely to cause detonation while using forced induction, it pays to use forged pistons and top-quality internal parts.

Superchargers

The supercharger is the most straightforward sort of forced induction. A supercharger is an air pump that is driven by a belt connected to the crankshaft. The pump forces air into the engine's intake system under pressure. Because it blows air into the engine, a supercharger is sometimes called a *blower*, although technically a blower is meant to

A supercharger on a 2008 Mustang. (*Photo courtesy Edelbrock.*)

move large volumes of air without a large rise in pressure. Using an actual blower for a supercharger normally won't make very much boost, unless you make substantial changes to the blower's design.

The amount of boost a supercharger puts out is determined by the design of the supercharger and the size of its drive pulleys. The drive pulleys determine how many times the supercharger turns for every turn of the crankshaft. There are several types of pumps that can be used as superchargers, but they can be divided into two large categories: positive displacement and variable displacement.

The word *positive* in *positive displacement* is used in much the same way as saying "I'm positive," to mean "I'm certain." A positive displacement supercharger will theoretically pump a constant amount of air for every time its input shaft makes a complete turn, no matter how fast the shaft is turned. The boost provided by this sort of supercharger is the same at every RPM, making it ideal for enthusiasts who want to build low RPM torque.

The most familiar sort of positive displacement supercharger is the lobe or Roots type. This is the sort seen on Top Fuel dragsters, Mad Max's Interceptor, and many American show cars. It uses a pair of rotors that spin in opposite directions to move air. Many of the superchargers used in early hot rods came off General Motors diesels, and are named after the engine series that such blowers were originally used on. The engine naming used the first number to indicate how many cylinders it had and the second number to give the cylinder size in cubic inches. For example, an 8-71 supercharger is patterned after the one on an 8-71 diesel, which was an eight-cylinder engine, with each cylinder measuring 71 cubic inches. Newer Roots-type superchargers, such as the ones from Eaton, usually use the amount of air moved in one input shaft revolution as a size number. While these superchargers are often the least expensive choice, they are also one of the least efficient. All superchargers require power to drive them, but a Roots type will take the most power from your engine.

On a V-type engine, a Roots-type supercharger sits on the top of the engine and often will increase the engine height. Depending on what kit you use and what car you

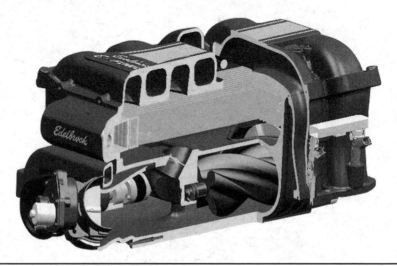

Cutaway version of a positive displacement supercharger. (*Photo courtesy Edelbrock.*)

have, installing a Roots blower may require cutting a hole in the hood. Whether this is a drawback or an added benefit depends on your sense of style.

There are several other sorts of positive displacement superchargers. The twin screw design, often sold under the Whipple brand, resembles a lobe-type supercharger but is considerably more efficient. If you go to shows that specialize in older British cars, you may run across a Judson vane–type supercharger. Aside from differences in mounting and efficiency, most positive displacement superchargers behave in the same way. They make a nearly constant amount of boost at every RPM.

Variable displacement superchargers produce more boost as the engine RPM increases. The only sort of variable displacement supercharger on the market today is the centrifugal type, which spins a rotor called an impeller inside a snail-shaped compressor housing. The centrifugal supercharger is highly efficient and requires significantly less power to turn it. The downside of a centrifugal supercharger is that it will not make very much boost, if any, at low RPM.

There are several ways to install a supercharger. Most Roots blower kits come with a new intake manifold, and the supercharger mounts to this manifold. Centrifugal superchargers usually mount to a bracket like an alternator or power steering pump and connect to the throttle body with a tube. Most supercharger kits offer relatively straightforward, bolt-on installation. Adjusting the level of boost requires switching drive pulleys to spin the supercharger faster.

All superchargers have two important characteristics. First, they must be sized to fit the engine. Most superchargers are at their most efficient at a certain range of RPM and airflow. A supercharger meant for a 1.8-liter Miata would overheat the air if you tried to use it to feed a 5.0 Mustang. By the same token, a blower meant for the 5.0 Mustang would be a poor choice for the Miata, even if you slowed it down to handle the smaller engine. A Roots-type blower would add a lot of unnecessary weight and friction, and would be less efficient from air leaking past the internal seals, while a centrifugal blower would hit its "surge limit" and deliver a rough, turbulent, noisy airflow that is likely to damage the supercharger. Second, superchargers require a significant amount of power to drive, sometimes more than 50 hp on a large engine making a lot of boost.

One quirk of superchargers is that the more freely your intake system flows and the more effective your camshaft, the less boost any particular supercharger setup will make. This is because a restrictive intake will create back pressure in the system that shows up on your boost gauge. If you add intake mods or a camshaft and find that you suddenly have increased the boost even though you have not tinkered with the blower, this is a sign that your engine is probably making less power than before.

AN ELECTRIC SUPERCHARGER?

Some "electric superchargers" have recently appeared on the market. This is a concept that could be workable, if the supercharger is as large as a conventional supercharger and driven by a sufficiently powerful electric motor. To give you an idea of how much power you need, delivering 5 psi of boost at 300 cubic feet per minute would need 5.6 kilowatts, or 7.6 hp, if you had a perfectly efficient compressor. Motors making that kind of power are usually about the size of a football. Such a motor would require a considerable amount of electrical upgrading, since you need around 400 amps to drive that sort of motor on 14 volts. Most alternators only put out 160 amps or less. An electric "supercharger" that was supposed to be a marine bilge pump is not likely to move enough air to make a difference. Worse, it may be so ineffective that it restricts your intake!

Turbochargers

There are more ways to turn an air pump than using a belt drive system. A turbocharger uses a turbine mounted in the exhaust system to spin a small centrifugal compressor. The extra restriction in the exhaust may rob the engine of a few horsepower, but it is considerably less than the power needed to drive a supercharger from the crankshaft.

Selecting the right turbocharger is not only a question of engine size, but a question of when you want the boost to kick in and how much boost you wish to run. A small turbocharger can spool up quickly to deliver boost at 2,000 RPM or less, but a larger turbo will be able to feed the engine more boost at high RPM for more peak power. Larger amounts of boost will not only require a larger turbo, but one designed to be efficient at a higher boost level.

Going to extremes can cause problems. A turbo that is too small may give a real kick off the line, but cause overheating issues if you hold the engine at high RPM, or even keep the engine from revving up to its maximum RPM. A turbo that is too large may not be able to make any boost at low RPM and act like an on/off switch at high RPM. It can also cause a delay between pressing the throttle and getting power, an effect known as turbo lag. Floor it in an engine with an oversized turbo, and the engine may wait a few

Turbocharged Datsun inline six. (*Photo courtesy DIYAutoTune.com.*)

A turbocharger. (*Photo courtesy Corky Bell/BEGi.*)

seconds and then have the turbo spool up with enough fury to smoke the tires. This kind of power may be impressive at a drag race, but can be annoying on the street, and downright treacherous in an autocross.

One method of getting more boost without the lag penalties is to substitute two smaller turbos for one larger one. These smaller turbos have lighter impellers and can spool up much faster. The downside of a twin-turbo setup is the expense and complexity required. Some factory twin-turbo engines have been known to use turbos of different sizes or valves that send exhaust through only one turbo at low RPM to further cut down on lag. This level of complexity is rare in aftermarket kits.

Turbochargers often need several other connections besides attaching the exhaust and the air sides. Most turbochargers require an external oil feed and oil return line. Water-cooled turbos also require a pair of water lines and often last longer than their waterless counterparts.

Although ball bearings have been around a long time, only recently have there been ball bearings capable of withstanding the red-hot temperatures and 100,000+ RPM found in turbochargers. Ball bearing turbos make a significant difference; the ball bearings produce less friction and can reduce the time it takes for a turbo to spool up.

Turbochargers often cause the exhaust manifolds or headers to become extremely hot. Some turbo setups use a bracket to support the weight of the turbo, but the heat

A tubular turbo manifold. (*Photo courtesy Corky Bell/BEGi.*)

still places stress on the exhaust system. Headers made of thin steel tubing often do not last very long on a turbo car, and are only a good idea on a race car where the headers can be inspected after every racing event or two. Stainless steel headers made from thin tubing can last a little bit longer, but are still often too flimsy for a car that sees regular street use without serious bracing. A durable manifold will need thick walls. A cast iron or cast steel manifold is often a very strong choice. If one of these is not available, a

Cast iron turbo manifold. (*Photo courtesy Edelbrock.*)

manifold welded from thick-walled piping will hold up better than one made from thin-walled tubing. The black iron pipe used for plumbing and sewer drains is much better for making durable turbo manifolds than thin exhaust tubing.

CHINACHARGERS

The number of inexpensive, Chinese-built speed parts has hit the turbo kit market harder than many other speed parts. Many of the parts are copied from existing American or Japanese kit designs, but many of the resellers often seem to have pieced together parts from several sources, and it's frequently unclear if the smaller resellers out there have ever tested out the kits they've pieced together on a real car.

I've seen a few of these kits work and make decent power, but these can be pretty dangerous territory for beginners. If you don't know what you're doing, the low cost of one of these kits can be a false economy. Here are some common complaints:

- **Incomplete kits.** It's rare for these kits to include an adequate tuning device, although reputable kits sometimes also will have this sold separately. Often, disreputable kits are missing a couple of important items like the downpipe (the pipe from the turbo to the rest of the exhaust) or oil lines for the turbo. Some of the worst offenders are little more than a turbo, an intercooler, and some aluminum tubing—you're on your own for figuring out what you will do about key parts like the exhaust manifold.

- **The turbos seem to be a grab bag.** I've seen some Chinese turbos rack up years of trouble-free street miles. Others start leaking oil after a month.

- **Most of the tubular exhaust manifolds on no-name kits tend to be prone to cracking.** I've heard of more complaints about the manifolds than the turbos. Cast iron manifolds haven't had as much trouble as ones made from tubing welded together.

- **Most of these kits come with absolutely no instructions or technical support.**

Of course, there are reputable turbo kits sold on eBay too, but they usually don't sell for $500 brand new.

Boost Controls

Controlling boost on a turbocharged car is a bit more complicated than controlling it with a supercharger. With no direct connection to the engine, the turbine gets its speed from the energy of the exhaust. The most reliable means of controlling a turbocharger is a wastegate.

A wastegate is a valve that opens to allow part of the exhaust gas to bypass the turbine. Opening the wastegate will slow down the turbocharger, while closing it will allow it to spin faster. An internal wastegate sits inside the turbine housing, while an external wastegate goes in the exhaust system upstream of the turbo. Many turbochargers come with an integral wastegate, but high boost levels may require the extra flow amounts that an external wastegate design can provide.

There are several ways to control a wastegate, and if you want to turn up the boost in an existing turbo setup, you will need some sort of adjustable boost control. A crude control would be to leave the wastegate stuck in one position and set that position so the turbo never exceeds the amount of boost desired. This system could work, but would probably only make maximum boost at one RPM point and cause the turbo to spool up slowly.

Most wastegate control systems use a pneumatic control. The simplest option is to run a line from the intake manifold to the wastegate actuator, a diaphragm mechanism that opens the wastegate. As the amount of boost rises, the wastegate opens further, and this limits the boost the turbo can make. To add control to this system, you can add an aquarium valve or similar small valve to bleed off pressure. Installing a small restrictor in the line downstream of the bleeder valve to force the air through a small opening such as a 0.050-inch orifice will keep pressure spikes from making the wastegate wildly flutter open and closed.

This system will control boost, allow for adjustability, and offer faster spool-up times than a fixed-position wastegate, but you can do better. For the fastest spool-up time possible, the wastegate should be kept completely closed below maximum boost and then opened as soon as the boost reaches its maximum. This can be accomplished by putting a mechanical valve in the air line that opens at the desired boost pressure and closes when the boost falls below this pressure. Such a system is fairly cheap and can even be built by a determined backyard enthusiast with parts from an industrial supply house.

A more sophisticated system uses electronics to control the wastegate, either by opening a valve in a pneumatic line or by operating an electronic actuator. These systems offer more adjustment than a mechanical control can provide. For example, you could make an electronic boost control add more boost at high RPM when power is starting to drop off, or use less boost in first gear for better traction. Electronic boost control can also compensate for different weather conditions. The ball valve boost control design tends to produce different levels of boost depending on air temperature.

If the throttle is located downstream of a turbocharger or supercharger, snapping the throttle closed suddenly can cause a pressure spike that can damage the compressor. The most common way to deal with this pressure spike is to install a valve near the throttle to release this spike. This valve is often called a compressor bypass valve, and it is designed to open only when you close the throttle while under boost. Not only will this valve protect your turbo, but it can help you keep it spooled between shifts. The most important feature to look for in a bypass valve is that it is able to handle the amount of boost pressure you are running without leaking. Very high-pressure systems may need a larger bypass valve to deal with the extra pressure.

There are two things you can do with the air that the bypass valve lets out. Some bypass valves send the escaping air to a spot upstream of the turbo, while others vent to the atmosphere. Venting the valve to the atmosphere can make a lot of noise and let everyone know you have boost, but the engine control unit (ECU) may not be able to account for the air escaping through the bypass valve. Most of the time, if your car has a fuel injection system that uses an MAF (mass air flow) sensor, you should not vent the bypass valve to the atmosphere without some way of tuning the ECU to handle this.

A compressor bypass valve that vents to the open air is sometimes also called a blow-off valve. Occasionally, you'll see the term "blow-off valve" used to describe a type of valve that opens at a specific pressure to protect the engine from too much boost, but this sort of valve is usually only something you'd see in racing events where the rulemakers want to prevent competitors from adding more boost. This sort of valve is also called a *pop-off valve*.

Intercoolers

Compressing air raises its temperature. This heat is not good for the engine. Not only does it decrease the air density, but it also makes the engine more likely to detonate. An intercooler can reduce this problem. The intercooler is a heat exchanger that cools the air as it exits a turbocharger or supercharger. Adding or upgrading an intercooler is a very good idea if you already have forced induction but want to turn up the boost. Some companies have been known to refer to their intercoolers as "aftercoolers," but the part is the same thing. Intercoolers come in two basic sorts: air to air and air to water.

An air-to-air intercooler looks a lot like a radiator. Usually, the best way to install one is to mount it behind the grill and in front of (or alongside) the radiator to draw in as much cold air as possible. When looking for an air-to-air intercooler, there are two main considerations: its cooling ability and its flow capability. The size is one of the biggest factors that determines its cooling ability. Adding small fins inside the intercooler tubes also boosts cooling ability. To find one with the best flow capacity, select one that uses many short tubes instead of a few long tubes. Tanks that do not make the air take any sharp bends also improve flow. If you have an intercooler with a 8" × 16" core, it is better to make this core by packing enough 8-inch tubes into a 16-inch-wide space than to pack a bunch of 16-inch tubes into an 8-inch-wide space.

Junkyards used to be a popular source of cheap intercoolers. Popular donors included Volvos, the Mitsubishi Starion/Chrysler Conquest, and the Ford Probe GT Turbo since the intercoolers on these cars often lend themselves to easy installation. Recently, cheap and serviceable intercoolers from China have flooded the market, putting an end to the days of paying $100 for a secondhand Starion intercooler for anything but a Starion restoration.

To work effectively, an air-to-air intercooler needs a place where incoming air will cool it down. Placing it in front of the radiator is almost always an effective location if

A lineup of intercoolers. (*Photo courtesy Corky Bell/BEGi.*)

you have the room. A duct to channel air through the intercooler can make this even more effective. Some cars like the Subaru WRX use a hood scoop to direct air to an intercooler mounted in the top of the engine compartment, while other cars locate the intercooler inside a fenderwell. Placing the intercooler behind the radiator, next to exhaust pipes, or any other place where it will be exposed to hot air is a formula for trouble. If the intercooler is in air that is hotter than the air leaving the turbo, it will act like an "interheater."

The other sort of intercooler is an air-to-water intercooler. An air-to-water system circulates a supply of cooling water from a reservoir to an intercooler mounted under the hood. The water then circulates to what amounts to a small radiator where it cools back down. Air-to-water intercoolers can go in the engine compartment or even be built into the intake manifold. These can work where an air-to-air intercooler would not be practical, such as mid-engined cars. Drag racers or autocrossers can also fill the reservoir with ice to give even more cooling for a short time. For a long race, however, this system is likely to absorb enough heat that it will become less efficient than an air-to-air system. Air-to-water intercoolers also suffer from added weight and complexity compared to an air-to-air intercooler.

Combining Forced Induction with Other Mods

Adding forced induction to an engine represents a major change. Some aftermarket mods will not even bolt up to the parts required for a turbo kit or supercharger setup. If you install a turbo, you will not be able to keep any aftermarket headers you may have installed, and probably not any cold air intake either. A supercharger may replace the intake manifold and cold air intake system. Of course, any decent kit will include its own replacements for these parts, designed to work with the rest of the kit. One of the other key changes is that an engine running forced induction is likely to need premium gas. Other than that, there are certain mods you can make to your engine to get the most from your boost.

Since you are cramming more air into the engine, your exhaust will need to be capable of getting all the air out. Exhaust mods for a supercharged car are often about the same that you would want to run on a naturally aspirated car making the same amount of horsepower. A good turbo kit will include an appropriate exhaust manifold or headers, but is likely to require a very high-flow cat-back or turbo-back exhaust. Since a turbo is not likely to be sensitive to oversized exhaust pipes, you can go all out here.

The fuel system and engine controls will need to be able to supply enough fuel to mix with the increased amount of air. At a minimum, a low boost kit on an injected engine may get by with an adjustable fuel pressure regulator and some means of retarding ignition timing. A better equipped kit will offer an electronic tuning device with either a ready-to-go tune or the ability to rewrite the fuel and spark maps as needed, and if necessary will come with oversized injectors or a higher-volume fuel pump. If no tuning method is provided, you'll need to get one elsewhere to make the car run properly.

Carbureted engines with forced induction usually work in one of two ways. A draw-through setup places the carburetor upstream of the supercharger or turbo. In most of these cases, the carb will need to be larger than the original and tuned specifically for this application. Although a draw-through carburetor can theoretically maintain a more accurate air-fuel mixture than a blow-through design, this design often suffers from slow throttle response and cannot use an intercooler. A blow-through setup, which is the most common seen in bolt-on kits today, leaves the carburetor where it was originally and feeds the boost through it. This approach requires either keeping the entire carburetor in a pressurized box or supplying boost to every opening on the carburetor that was originally open to the atmosphere. In either case, it is best to have an expert tuner adjust and modify the carburetor to work with your engine.

ECU mods will usually adjust the spark timing along with the fuel system. If you have a mechanical ignition, the distributor will need to be recurved. Installing ignition upgrades if you have a distributor is also often a good idea because the higher pressure in the cylinders may require a hotter spark to ignite. Some newer distributorless ignitions are hot enough that there's no need to upgrade, only to adjust the timing.

Since the higher pressures from forced induction can cause an engine to detonate, lowering the compression with different pistons is often a good idea if you started with a high-compression motor. Some turbo engines have a compression ratio of 8.5:1 or even lower, compared to over 10:1 for many high-performance, naturally aspirated motors. Running forged pistons for extra strength is also a good idea if you are running high levels of boost or your stock pistons have a reputation for being flimsy. For further toughness, you can add an extra-strong aftermarket head gasket sealed down with stronger cylinder head bolts or studs.

Forced induction also benefits from removing restriction in other areas. Oversized throttle bodies, cylinder head porting, and oversized valves will all free up more power. These mods can make even more of a difference with boost than without it.

Turbochargers and superchargers can also affect your cam choice. Turbochargers in particular do not work well with cams designed for high-revving, naturally aspirated engines. The pressure in the exhaust manifold on a turbo motor is often higher than the boost pressure, and if the exhaust valve is open enough during the intake phase, the exhaust will flow backward into the cylinders. For a turbo motor, short duration and small amounts of overlap will often be the best choice for your camshaft. Newer engines often work best with nearly stock camshaft grinds, while on an older engine, it may be best to use a "towing" or "RV" cam if you cannot find one made specifically for turbos.

Nitrous Oxide

Turbochargers and superchargers feed the engine more air by pulling in more air from the atmosphere. But what if you could just add oxygen straight from a bottle? This may sound like a good idea, except that pressurizing your intake with extra oxygen might cause the fuel to ignite far too early, maybe even before it could get into the cylinders. The solution to this is to use something a bit more stable: nitrous oxide.

Nitrous oxide, or N_2O, is a chemical containing an oxygen atom attached to a pair of nitrogen atoms. At high temperatures, it splits into nitrogen and oxygen, leaving the oxygen free to combine with the fuel once the mixture has already ignited. Since this is not a reaction that happens at room temperature, nitrous oxide is a much safer oxidizer to use than pure oxygen. Nitrous oxide is also used as laughing gas, but the industrial formula used in cars has a few additives put in that would make it considerably less pleasurable to use as a recreational drug. Besides laughing gas, nitrous oxide is also frequently referred to as the juice, the bottle, or simply as nitrous. While NOS is a brand name of nitrous oxide injection, few serious tuners will pronounce it "naws" unless they are making an ironic movie reference.

Direct port nitrous on an intake manifold. (*Photo courtesy Edelbrock.*)

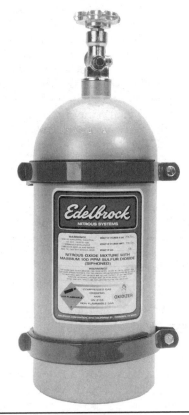

A tank of nitrous oxide. (*Photo courtesy Edelbrock.*)

A nitrous system must also supply enough extra fuel to react with the nitrous. Leaning out a bottle-fed engine is likely to cause severe detonation, so it is safer to run too rich than too lean. Always follow the manufacturer's instructions for how to tune a system if it can be set to make different levels of power.

There are several sorts of ways to deliver nitrous oxide. A dry-flow nitrous system, which is really only practical with fuel injection, uses a sprayer in the throttle body to deliver nitrous oxide and supplies extra fuel by either increasing fuel pressure or holding the fuel injectors open longer through electronics. Wet-flow systems add both the nitrous and the fuel at a single point near the throttle. Many carbureted engines use a plate-type system, a sort of wet-flow system where a plate below the carburetor contains a device called a spray bar that sprays out a mixture of nitrous oxide and gasoline. Port-type nitrous systems have individual nozzles at each port that add a nitrous-fuel mixture. Port-type systems can often make more power due to better fuel distribution. Dry-flow systems can also have excellent fuel distribution if they control how long the injectors open. Dry-flow systems that modify fuel pressure are less accurate at their air-fuel mixing, giving less fuel at low RPM than at high RPM. Most systems use solenoid valves—fast-opening valves opened electronically—to turn the nitrous and fuel on or off.

Most nitrous oxide kits are rated by the amount of horsepower they can theoretically add. This is determined by the amount of nitrous and fuel supplied. In theory, a 100-hp shot of nitrous would boost horsepower by 100 hp on any engine, at any RPM. This means that nitrous adds far more torque at low RPM than high RPM. A common rule of thumb is that without any internal modifications, it is safe to add a 50-hp shot to a four-

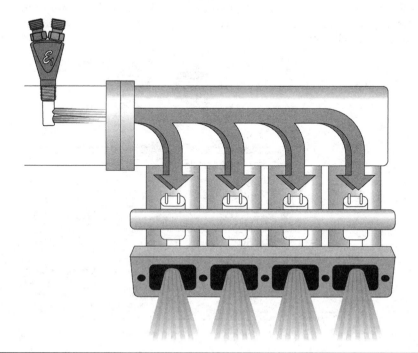

Wet-flow nitrous on a fuel-injected engine. (*Illustration courtesy Edelbrock.*)

Nitrous spray bar for a carburetor. (*Photo courtesy Edelbrock.*)

cylinder engine, 75 hp to a six, and V8s can add 100 hp. A motor with the internal mods to take the higher cylinder pressures and the risk of detonation can handle more. The internal mods for nitrous are often close to the same ones for other forms of forced induction.

Some turbocharged cars use a small shot of nitrous, around 50 hp or so. This can often provide more benefit than the small rating would imply. Since nitrous drops to subzero temperatures as it exits a sprayer, nitrous acts like an additional intercooler. Also, since nitrous makes power even at very low RPM, it can help an oversized turbo spool up much faster.

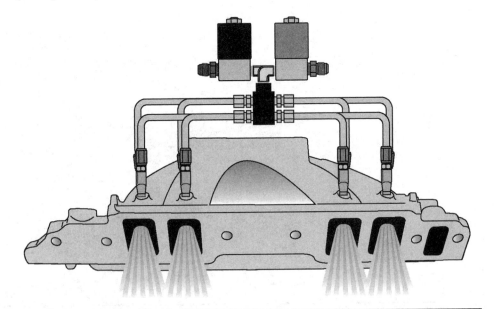

Direct port nitrous. (*Illustration courtesy Edelbrock.*)

There are several accessories one can add to a nitrous system. The most visible is a purge system that opens small valves in the lines to let in fresh nitrous and let any air that was in the lines escape. Purge systems often let out visible clouds of escaping nitrous, which is often a popular crowd-pleaser at drag races. A bottle warmer is useful for keeping the bottle pressure from dropping in cold weather. There are several control boxes, such as "window switches" that only allow nitrous to be supplied between a certain minimum and maximum RPM, and dual-stage nitrous systems that let you start a race with a small amount of nitrous and then add more power as soon as you are moving fast enough to do so without smoking the tires. Progressive nitrous systems can accomplish a similar goal by opening the solenoids slowly on a time delay.

Nitrous oxide is an extremely cheap way to add power to an engine, but it is not without its drawbacks. Most systems have an on/off nature, making them a poor choice for tackling a winding road. The bottle seldom lasts for more than a minute or so, making this mostly suitable for when you need a short burst of speed. If this were not enough to restrict it to drag racing, most other sorts of racing do not allow it at all. Nitrous's outlaw reputation has also caused the police to take a dim view of anyone they pull over with a system installed. In some states, driving with a nitrous system is not allowed on public roads at all unless the bottle is disconnected.

Deciding Which One You Need

When it comes to forced induction, all three choices have their own strengths and weaknesses. Which one to pick depends on your goals. There is no "best" power adder, only ones that are best for certain purposes.

Nitrous lets you get a burst of power with a minimal initial expense. It can cost more in the long run, however, since filling a ten-pound bottle and paying three to four dollars a pound can add up. Nitrous is a reasonable choice for drag racing.

Turbochargers allow for enormous power gains, but are usually the most difficult power adder to install. They often require a very large number of connections and require intricate controls. However, many enthusiasts find they are worth the trouble for the power gains and tuneability. A turbo running at relatively low boost can work quite well for almost any application, from street driving to drag racing to autocross. Unfortunately, at higher boost pressures, turbos start to take on an on/off nature that makes the car tricky to drive except for drag racing. When it comes to drag racing, though, turbos tend to dominate almost any class where the rules allow, even if the rules allow competitors with other power adders to run larger motors or lighter cars.

Superchargers offer relatively easy installation, good performance, and good street manners for most kits. A positive displacement supercharger is a great choice for torque lovers, but their lack of efficiency limits how much boost you can get from one, and they typically make less power for a given boost level. Centrifugal superchargers tend to come alive at high RPM and may result in a car that is not as fast for the same amount of boost or horsepower. On the other hand, their efficiency means that they can make more power reliably than positive displacement designs. Centrifugal superchargers are also frequently the easiest to install in a cramped engine compartment.

Which power adder is right for you depends on your priorities.

CHAPTER 7

Engine Tuning

Adjusting the Engine Components and Controls for Best Performance

Sometimes, choosing and installing the right engine mods is only half the effort needed to make your car go faster. The rest of the effort lies with tuning your engine, adjusting it to add just the right amount of fuel and fire the spark plugs at precisely the right moment. Some mods do not require very much tuning. If you have added some basic intake and exhaust mods to a fuel-injected engine, the engine control unit (ECU) may simply be able to detect that the engine is pulling in more air and adjust to compensate. More complicated mods may confuse the computer. If your engine uses a carburetor and mechanical spark controls, even small changes like adding headers can call for tuning to get the most out of your mods.

Tuning is not just important for power, as poor tuning can shorten the life of the engine. One of the most serious problems that can happen when the controls get the tuning wrong is detonation, also known as pre-ignition, spark knock, or pinging. Pinging occurs when the fuel ignites before the spark plug should light it. In this case, the fuel explodes instead of burning with a controlled flame. This produces a rattling sound that may sound like rocks in a coffee can or a sharp "ting" sound depending on the engine, and produces shock waves that can damage the pistons. Pinging can be caused by running the fuel too lean (not enough fuel for the amount of air the engine has pulled in), by advancing the spark timing too far, by running too low an octane of gasoline, by too much compression, or by too much boost. If you ever hear detonation's distinctive rattling after tuning or installing a mod, lift off the gas immediately, and fix the tuning as soon as reasonably practical.

Detonation is not the only peril you may encounter when tuning an engine. If you add far too much fuel, it will not all burn. The unburned fuel can slip past the piston rings, washing oil off the cylinder walls and diluting the oil. So adding too much fuel can make the engine wear out faster. Other cases of bad tuning may cause an engine to run too hot. Luckily, this problem is one you can easily spot by keeping an eye on the temperature gauge.

Tuning Greg Amy's Nissan NX2000 to get ready for a race. (*Photo courtesy DIYAutoTune.com.*)

Telling Which Adjustments the Engine Needs

If you were reading a set of directions and came to a line that just said, "Turn this knob," your first question would probably be, "Which way am I supposed to turn it?" When tuning your engine, you will need to figure out if you need more fuel or less, and more spark advance or less. Sometimes there is an easy way to tell. For example, if you put on a set of fuel injectors that flow twice as much fuel, you will need to set them to open for only half as long. Usually, however, you will need other ways of checking.

The air-to-fuel ratio can be the simplest one to set, if you have the right tools. This is usually measured in terms of pounds of air to pounds of fuel. If you have 14.7 lb of air for 1 lb of gasoline, the air will theoretically supply enough oxygen to burn all the fuel, without any oxygen or any fuel left over. This is called the *stoichiometric ratio*, from a chemistry term that means you have the right amount of chemicals present. If you add more fuel and bring the air/fuel ratio below 14.7:1, the mixture is described as rich. Reducing the fuel so the ratio goes above 14.7:1 results in what is called a lean mixture. The best mixture to use depends on what the engine is doing. An engine will often make its best power with a slightly rich air/fuel ratio, around 12.8:1 to 13.5:1. Turbocharged or supercharged engines often need to run richer, from 11.6:1 to 12.0:1, to avoid detonation under heavy boost. The best gas mileage occurs when the engine is running an air/fuel ratio in the range of 15.5:1 to 16.5:1. Engines typically produce the least pollution running at around 13.5:1 to 14.7:1. (Note that some gasoline types have slightly different ratios, depending on the additives. Some blends may have 14.6:1 or 14.5:1 as their stoichiometric air/fuel ratio.)

These ratios provide several useful targets. With the engine running flat-out at full throttle, you will want to adjust the mixture to the rich end of the operating range. Cold starts are another area where engines tend to require a richer mixture. When cruising at part throttle at the usual cruising RPM, leaning out the mixture to as low as 16.5:1 can improve gas mileage. If you tune the engine to run this lean at cruise, make sure it goes back to a richer mixture when you open the throttle. To pass an emissions test, you may want to bring the mixture closer to stoichiometric. If you adjust the engine to run leaner under load and it starts to ping, adjust it to run richer as soon as you can.

Knowing where to set the mixture is not especially helpful if you have no way of knowing what mixture you are running. The most direct method is to measure the mixture with a wide-band oxygen sensor. This handy device goes into your exhaust system and can directly measure almost any air/fuel ratio that your engine can run. Unfortunately, the price of a wide-band oxygen sensor and the electronics to control it starts at around $200. They are very useful if you can afford them. If you are having your car tuned on a chassis dyno, the dyno shop will probably have one that they will temporarily attach.

There are several other methods of estimating your air/fuel ratio. Most fuel-injected cars are equipped with narrow-band oxygen sensors, and you can add one if necessary. This sensor normally works with the ECU directly, but you can add an air/fuel ratio gauge to watch its readings for yourself. These sensors only show useful data when the air/fuel ratio is close to stoichiometric. Beyond that, they can tell you little more than if your mixture is too rich or too lean. This may not help you find where you can make the best power, but it can keep your emissions on target and prevent a naturally aspirated engine from running dangerously lean.

Shade-tree mechanics have long relied on another method: reading the spark plugs. To use this, scrape the spark plugs clean of any deposits and then make a few full-throttle runs, preferably someplace where this is legal like on a dragstrip. Then remove

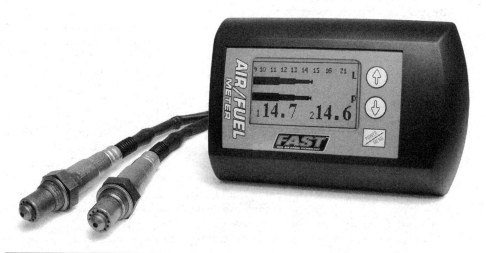

A wide-band air/fuel meter with two sensors. (*Photo courtesy Comp Performance Group/FAST.*)

the spark plugs and look at the color of the metal electrodes. They should be brownish, tan, or reddish. If they are black and sooty-looking, your mixture is too rich. If they have turned white, your engine is running too lean. A definite danger sign is if there are silver or scattered black spots all over the tip, including on the porcelain insulator. These happen when detonation causes small flecks of aluminum to melt off the pistons. Another sign of trouble is if the plug tips are covered with oil, which is a sign your engine is starting to wear out and let oil into the cylinders. Reading spark plugs is often useful as a repair technique, but it can give a bit of help at tuning too. It's especially useful for spotting if one cylinder has a problem compared to the others.

While the air/fuel ratio is a useful number to know, horsepower and quarter-mile times are often the most effective ways to judge if you have tuned your engine correctly. Although you may want to get your engine somewhat dialed in before putting it on a dyno, measuring horsepower and using trial and error can often get the most power out of your combination. If you cannot afford time on a dynamometer, testing at a dragstrip can be the next best thing.

While dyno or dragstrip testing is often the finishing touch for adjusting the air/fuel ratio, these methods are often the only ones that work reasonably well for finding the perfect spark advance curve. The usual method is to adjust the timing with considerably less advance than expected and then to gradually advance it until the engine makes less power, at which point you back it off to the spot where it made the most power. Occasionally an engine with a turbo or supercharger may start pinging before you reach the advance needed for maximum power, in which case you should either back the timing off a few degrees from the edge of detonation or run a higher-octane fuel.

There are, however, a few basic rules one can follow for adjusting the timing. Timing is usually determined primarily by engine RPM and the level of vacuum (or pressure, if you are running forced induction) in the intake manifold. An engine will need more spark advance as the RPM increases. Heavily modified engines often need to advance the timing faster than lightly modified ones. At most of the high RPM range, the engine will need almost the same amount of advance through a wide section of RPM. If you have computer-controlled timing, you can often widen out the powerband a little by adding more timing after the torque peak. This is next to impossible with a mechanical distributor advance, however.

When cruising at part throttle, the manifold vacuum increases, and you can safely advance the timing even more. The opposite rule is that if the pressure in the manifold increases, you must retard the timing to prevent detonation. Many engines from the 1970s used a connection called ported vacuum to avoid applying the vacuum advance at idle. The idea behind this was to warm up the catalytic converter faster, but like a lot of early attempts at emissions control, it wasn't very good for drivability.

Since you can advance the timing if the pressure in the intake manifold drops, you also need to retard the timing when the pressure rises. If you are using forced induction, the ignition system will have to reduce the timing advance as the boost rises. The higher pressures from forced induction make an engine more likely to detonate unless you adjust the timing to compensate. There are a few versions of vacuum-advance mechanisms that also retard timing under boost, but these are not easy to find. MSD and several other electronics companies make devices that can adjust timing under boost electronically. Although nitrous oxide does not raise intake manifold pressure, these systems do require retarding the timing when in use. Nitrous timing retard is something you usually need electronics to handle if you don't want to just tune your ignition for good performance when spraying nitrous.

Tuning Computers

In a way, computerized controls can be easier to adjust than their mechanical counterparts. On a car like this, the ECU simply reads the signals from several sensors on the engine and decides when to add fuel and fire the spark. Retuning the computer is simply a matter of telling the computer a new set of rules for how to respond to the sensors.

Original ECUs do not always need adjustment. To some degree, these computers can make minor adjustments to the amount of fuel added to compensate for mods installed. Most factory ECUs can work with minor bolt-on mods such as headers or cold air intakes. The more mods you add, however, the further you push the computer into territory that the original engineers did not expect the computer to see. In fact, if your car is equipped with OBD-II, the computer may conclude that something is broken if you add too many mods.

The OBD (on-board diagnostics) system examines the various sensors located throughout the engine compartment to determine if the engine is running correctly, to make sure the emissions are within specs, and to help mechanics diagnose problems. If the system detects any problems, it will set off the check engine light and allow a mechanic to read the information about what is wrong using a scan tool.

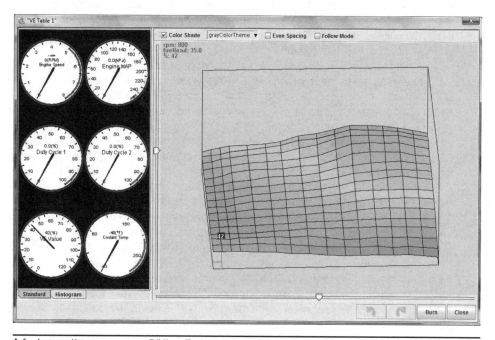

A fuel map. (*Image courtesy DIYAutoTune.com.*)

THE ECU AND ITS SENSORS

An ECU needs to measure many things about an engine. At a minimum, one controlling the fuel injection will need to know how much air is coming into the engine. There are three common methods an ECU can use to figure out how much air is coming into an engine: MAF, speed density, and Alpha-N.

MAF stands for mass air flow. An MAF system uses a sensor to directly measure how much air is entering the engine. The most common MAF sensor uses a heated wire that the incoming air cools. The computer can determine the amount of incoming air by the temperature of the wire. Some other MAF systems use a vane airflow meter, where the incoming air pushes a spring-loaded trap door open. MAF sensors are only accurate through a limited range of airflow levels, but they work very well at compensating for wear or other changes to the engine.

Speed density measures how fast the engine turns and the pressure of the air in the intake manifold. The pressure is measured with a MAP, or manifold absolute pressure, sensor. The computer takes the measurements of the engine RPM and the MAP sensor reading, compares these to a chart in its memory, and uses the chart to determine how much fuel to add. Speed-density systems are much more likely to need retuning if you add internal engine mods, but they can sometimes handle a wider range of airflow than an MAF sensor. This makes speed density a good choice for heavily modified turbo cars. Most aftermarket controllers use speed density.

Occasionally, you may see a fuel injection controller that uses Alpha-N. This system determines how much fuel to add using only the throttle angle and engine RPM. Some naturally aspirated racing engines use Alpha-N to avoid problems with unsteady air flow caused by wild camshafts. A few inexpensive aftermarket systems such as the Holley Pro-Jection have exclusively used Alpha-N as well. Alpha-N can work acceptably on naturally aspirated engines, but is difficult to tune on a supercharged engine, and almost completely unworkable on a turbocharged one.

Normal fuel injection systems use many sensors besides the MAP or MAF sensor. When you suddenly open the throttle, the injection system needs to add more fuel. Consequently, even systems that do not use Alpha-N often have a throttle position sensor. The ECU also needs data on the temperature of the incoming air, so unless the MAF has a built-in means of measuring air temperature, the engine will need a temperature sensor somewhere in the air intake. Most engines also have a second sensor monitoring the coolant temperature. The ECU also needs to monitor the crankshaft as it turns. What sort of sensors it needs depends on whether it uses a sequential or batch fire system.

The ECU can either time the injectors to fire based on when the intake valve opens and closes, or simply spray enough fuel into the intake without paying attention to the valves. The first system is known as sequential fuel injection, while the second is typically known as batch fire because it activates the injectors in batches. Surprisingly, sequential fuel injection systems do not give a significant horsepower difference unless you take the time to tune each cylinder individually. With normal fuel injectors, most sequential fuel injection systems have to switch to batch fire at high RPM. The injectors simply cannot open and close fast enough to work in a sequential system through the entire RPM range. Sequential injection is common on original equipment systems, and has recently become fairly common in aftermarket ECUs.

Sequential systems need a precise way to determine when to fire the fuel injectors. Engines using this method need a crank angle sensor and a cam angle sensor. An angle sensor typically looks like a gear with as many as 60 teeth and 1 or 2 teeth missing. A magnetic or optical sensor sits next to the wheel and detects the teeth as the wheel turns. Batch fire systems can get by with a less precise system, usually a trigger wheel in the distributor with as many teeth as the engine has cylinders.

The ECU also uses oxygen sensors to measure the air/fuel ratio. When cruising, the ECU may go into what is known as open loop mode, where it determines how much fuel to add based on oxygen sensor readings. Open loop mode generally does not work well at full throttle, where the engine needs a richer mixture than a narrow-band oxygen sensor can measure.

Which system your car has can have an effect on how the car responds to mods. OBD-I systems built in the late '80s to early '90s often allow considerably more changes, even removing the catalytic converter, before setting off a check engine light. The OBD-II computers that appeared in the 1996 model year are much more likely to mistake a mod for a breakdown. If an OBD-II computer detects readings that its programming says should not be happening, it may ignore sensor readings or go into a "limp-home mode" that drastically cuts power. If your mods set off a check engine light, you may need to reprogram the computer or replace it with an aftermarket ECU.

If your ECU's tuning no longer matches your engine, there are several ways to fix this. Some factory computers can be retuned with chips or plug-in programming devices, depending on the design of the computer. At the other end of the spectrum, you can replace the original ECU with an aftermarket module that allows you or a dyno operator to tune the computer. Piggyback computers and black boxes fall somewhere in between. These devices make changes to the signals the computer receives to "trick" the computer into changing fuel, or sometimes ignition, settings.

Performance chips gained popularity in the '80s and are still a viable option for some cars. Most chips work with OBD-I and earlier ECUs. These may come as a plain electrical component that must be snapped into the original ECU, but sometimes a chip will arrive in a sealed ECU that replaces the original unit. Most off-the-shelf chips are designed to work with stock engines or ones with minor mods. The gain from one of these is likely to be only a few horsepower, except on turbocharged cars, which have a computer that controls the boost level or older cars with less well-designed programs.

On an engine with internal work, you will need a chip designed to match your mods. A chip designed for a different set of mods may not be what your engine needs. Consequently, you may need to have a chip custom made unless you run a common package of mods.

On some popular car brands, particularly General Motors and Honda products with OBD-I or earlier computers, it is possible to program your own chips. These computers store their tuning data in an EPROM (erasable programmable read-only memory) chip. This chip can be removed from the computer, erased, and reprogrammed with a device called an EPROM burner before reinstalling it back into the computer. Doing this will require a PC, an EPROM burner from an electrical supply company, and software that can be downloaded from the Internet. This approach can be so effective that some owners have adapted GM computers to cars from other brands. More information on making your own EPROM chips can be found at Internet communities dedicated to ECU reprogramming. This may not be the most user-friendly approach to tuning, but it can be done on a very low budget.

A plug-in programmer. (*Photo courtesy Edelbrock.*)

Plug-in programming devices usually work with OBD-II computers and come in two versions. One has a single built-in program and works in much the same way as a performance chip. The other sort is more expensive and allows for making custom changes. In some cases, it is possible to download software for a laptop computer and connect it to the ECU with an adapter cable.

Piggyback controllers work along with the original ECU. They change the signals the ECU receives from its sensors to make the ECU adjust the fuel or spark timing. Most piggyback systems will let you start with the original settings and gradually move things from there. These setups are inexpensive and can present fewer tuning difficulties than working with a replacement ECU. On the downside, a piggyback controller has a smaller range of adjustment than a new ECU. Many of them only allow adjusting fuel and do not permit adjusting the spark advance.

In addition to piggyback tuning controls, there are several add-on "black boxes" built for single purposes. Some are designed to bypass built-in safeguards such as speed limiters. Others control a single aspect of the engine. For example, on Honda engines it is possible to buy a VTEC controller to change the point at which the VTEC variable valve timing system turns on.

A stand-alone aftermarket ECU is often the most expensive solution, and the most work. For this cost, you get the most versatility in tuning the engine. Many of these systems cost over $1,000, compared to around $300 for some piggyback devices. Stand-alone systems can allow complete control over every function provided by a normal ECU. An aftermarket ECU is a good choice for heavily modified engines that have larger injectors, forced induction on an engine not originally designed for it, or extreme internal work. The trade-offs, besides expense, are the difficulty of tuning one and the fact that many of these lack the sophisticated diagnostics that a stock ECU from the past decade will have. (If your ECU is from the '80s or early '90s, on the other hand, you may get more diagnostic information out of a stand-alone aftermarket ECU than the stock unit.)

Stand-alone ECUs come with several different levels of features. The most minimal designs offer complete control of the fuel injection and leave spark control to a mechanical distributor or separate spark controller. Most stand-alone systems also offer spark control, and high-end models even offer extra outputs for controlling nitrous oxide, turbo pressure, or virtually any other device you might want an engine control to activate.

There are many features to look for on an aftermarket ECU that can make life considerably easier when installing it. A plug and play ECU is designed so that it will connect to your engine using the same connectors as the original ECU. Such an ECU can be swapped out just as easily as swapping in a new stock computer. If an ECU is not plug and play, you will want to check if it can work with your original sensors. If not, you will need to install new sensors that match the sort the ECU was designed to work with.

At the other end of the difficulty scale from plug and play is the original MegaSquirt ECU. This can be bought for less than $300, but the catch is that it arrives in several small bags of resistors, chips, and other electrical components. You will have to solder it together yourself at this price point, and making the wiring harness to connect it under the hood is also your own responsibility. This ECU was originally designed strictly as a fuel control with no spark control possible, but later developers have added a variety of spark control systems and all the features you'd expect of a commercial ECU. While the classic MegaSquirt was a solder-it-yourself kit that was as much an educational tool as a tuning tool, distributors now offer ready to run, and even completely plug and play, versions of the MegaSquirt.

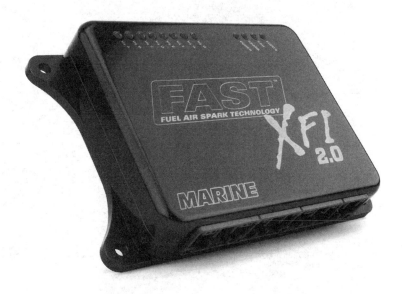

A stand-alone ECU. (*Photo courtesy Comp Performance Group/FAST.*)

Carburetor Tuning

At a glance, a carburetor may look simpler than a fuel injection system. You simply run a fuel line to one small part that takes care of all the engine's fuel mixing needs without any external sensors or other complications. A closer look at a carburetor, however, can give you the impression that this one part has squeezed all the complexity of a fuel injection system into a box the size of a paperback Tom Clancy book. The typical carburetor contains an assortment of pumps, valves, moving needles, and tiny tubes running every which way.

The core of a carburetor is the venturi, sometimes called a "barrel." This is the tube that the incoming air enters, where the carburetor adds fuel. The venturi narrows down in its center, often with small circular objects called boosters to constrict it further. The throttle plate lies at the bottom of the venturi. As air flows into the venturi, it must speed up as it travels through the narrow section. As the air speeds up, its pressure drops. A fuel tube sticks out into the middle of the venturi, where the low-pressure area sucks fuel out of the tube and into the incoming air. Most of the rest of carburetor design aims to make sure that the right amount of fuel comes out of the tube. Carburetors may have from one to four venturis.

The key to tuning a carburetor is vacuum signal. The vacuum in the venturi becomes stronger the faster the air flows through it. If the air flows through a venturi too slowly, the carburetor can no longer add fuel accurately. This is why it is important to select a carburetor that is not too large for your engine. Some manifolds change the vacuum

Adjusting a carburetor. (*Photo courtesy Edelbrock.*)

signal. For example, a dual-plane manifold often creates a stronger vacuum signal and can allow using a larger carburetor than calculations would recommend. In many cases, installing a different manifold requires adjusting the carburetor to run with the different vacuum level.

The tube gets its fuel from a float bowl. This chamber at the end of the carburetor gets its name from the float assembly that controls the level of fuel in the chamber, in much the same way as the float in a toilet tank controls the water level. The fuel enters the float bowl through a needle valve. When the fuel level is low, the float lets the needle valve open to let in more fuel. As the float rises, it pushes the valve closed. If the float is adjusted too high, gasoline may spill into the carburetor through the wrong tubes. If the float level is too low, the fuel may slosh around and leave the openings that it should enter high and dry. Some carburetors have parts like sight glasses to help you tell if the level is correct.

Floats come in several designs. On four-barrel Holleys and other popular carburetor designs, you can buy different float bowls to fit different types of racing. Different designs change how acceleration, braking, and cornering affect the accuracy of the float. Center-hung floats are not affected by cornering hard, while side-hung floats operate better under extreme acceleration or braking.

A part called the main jet restricts the flow of fuel from the float bowl to the venturi. The simplest sort of jet is a screw-in plug with a precisely drilled hole through its center. Putting in a jet with a larger hole makes for a richer mixture, while one with a smaller hole will make the mixture leaner. Replacing a jet often requires disassembling the float bowl, although some aftermarket kits allow changing jets without this. Percy's Adjust-

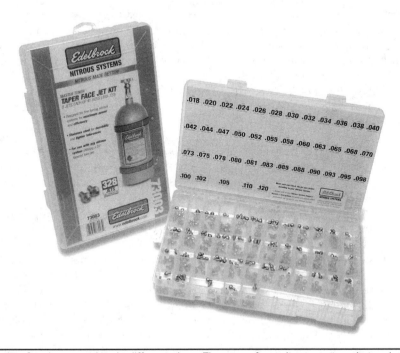

A package of replacement jets in different sizes. These are for a nitrous system, but carburetor jets are almost identical. (*Photo courtesy Edelbrock.*)

A-Jet kit replaces the main jet with a needle valve that you can adjust from outside the carburetor. Other kits simply allow pulling out the jets without disassembling the entire float bowl.

The main jet can supply a relatively constant air/fuel ratio over much of the engine's RPM range, except near idle speeds. Unfortunately, the engine does not work best with a relatively constant air/fuel ratio. Hard acceleration, cruising, and warming up the engine all call for different mixtures. Carburetors typically contain four extra devices for mixture control: the power valve, the accelerator pump, the choke, and the idle circuit.

The power valve is a valve controlled by manifold vacuum, the vacuum level downstream of the carburetor. At part throttle, the manifold vacuum is high and the power valve is closed. At full throttle, the vacuum level decreases and the valve opens. The power valve allows fuel to flow alongside the main jet, making the carburetor behave as if it had a larger-sized main jet. This allows a richer mixture at full throttle and a leaner mixture for cruising. To select a power valve, measure the manifold vacuum at full throttle and maximum RPM at the dragstrip using a vacuum gauge. The power valve should be open at that vacuum level and with just a little more vacuum. If it opens with substantially less, you'll use more fuel in light throttle cruising.

Some carburetors do not use a power valve, but instead use a vacuum-controlled needle that moves in and out of the main jet. This needle results in a leaner mixture when there is more manifold vacuum. Different needles are available to tune for different engines. Carburetors that use needles include the Edelbrock, Carter four-barrel, and SU carburetors found on many British cars.

When you open the throttle suddenly, the engine needs more fuel than the main circuit supplies. The accelerator pump supplies this extra fuel. Different carburetor designs use different methods to connect the accelerator pump, but they all are designed so that opening the throttle causes the pump to deliver extra gasoline. The engine will "bog" or hesitate when you step on the throttle if the accelerator pump is not adjusted correctly. On the common four-barrel Holley, you can control the accelerator pump's behavior by changing a cam (in this case, a curved bit of plastic) on the throttle shaft to alter the amount of fuel and the point during the throttle opening at which the pump adds it. Further control is possible by replacing the nozzle or by converting it to use a larger pump.

The best way to tune the accelerator pump is to adjust it so that it supplies just enough fuel to avoid any hesitation or flat spots in the acceleration. The best way to find the right tuning is to start with the setting that adds as much fuel as possible and then gradually try settings that add less fuel. Once you have reached a point where you can feel any hesitation, go back one step.

The idle circuit is a series of passages in the carburetor that supply fuel at idle and part throttle. Usually, you can control the idle speed with an adjustment screw that sets where the throttle stops when closed. Most carburetors also have a screw that allows you to adjust the air/fuel ratio at idle. The fast idle adjustment is a second stop on the throttle that only operates when the engine is cold, causing it to idle at a higher RPM.

The choke is an extra throttle-like valve at the top of the carburetor. When the engine is cold, the choke closes, making the mixture richer. Chokes may be controlled by a mechanical thermostat, an electrical device, or a hand-operated cable known as a manual choke. The engine needs to have the choke open to make maximum power, unless the tuning is very wrong. Some racing carburetors omit the choke because it restricts airflow, with the trade-off that the car is harder to start in cold weather.

Carburetors often contain several of each system. For example, a four-barrel Holley Double Pumper contains four venturis, four main jets, two power valves, and two accelerator pumps. The advantage of this extra complexity is that it can use two barrels for low-speed driving and then open all throttles for full power. This gives it accuracy over a wider range of speeds than an equally sized one-barrel carburetor.

With four-barrel carburetors, you have a choice of methods to open the secondaries, the two throttles that open only for full power. Vacuum secondaries and their cousin, the air valve secondaries, are designed to open at full throttle when the engine RPM becomes high enough to need all four barrels. These provide good drivability and can be a good choice for street use. Mechanical secondaries usually open the secondary throttles after the main ones are partially open, and require a second accelerator pump. Synchronous opening mechanical secondaries open at the exact same time as the primaries and usually show up only on cars built strictly for drag racing.

Spark Timing

Ignition systems come in two main categories: distributors and multiple coils. Distributors may control the timing with a mechanical system mounted in the distributor or by an external computer, while systems with multiple coils require computer control. Multiple coil systems include several different designs, such as wasted spark systems that send a spark to two cylinders at once using a coil with two terminals. This spark happens in the exhaust stroke, so the second spark does nothing. Another sort is known as coil-on-plug, where the coils attach directly to the spark plugs with no wires.

When it comes to adjusting the spark timing, computer-controlled systems all work in much the same way. A computer takes a signal from a sensor in the distributor, on the camshaft, or on the crankshaft. The computer calculates the timing needs based on engine load, RPM, and sometimes coolant temperature or other factors. The computer then uses the signal from the external sensors to determine when to fire the spark plugs. Tuning a computer-controlled ignition usually requires reprogramming the stock computer or using a programmable aftermarket ECU with spark control. A few high-end piggyback controllers can also change the timing curve, but most piggyback systems control only the air/fuel ratio.

The easiest adjustment to most distributors is to rotate the distributor clockwise or counterclockwise. Many distributors have a hold-down screw that you can loosen to permit changing the timing this way. Turning the distributor will advance or retard every point in the spark advance curve by the same amount. On many older cars, advancing the timing can make a real improvement. To get the absolute best advance curve possible, however, you will need to modify the distributor's internal mechanisms to tweak the timing at different points.

There are two parts to a mechanical spark control on a typical street car. The centrifugal advance mechanism causes timing to advance as RPM increases. The vacuum advance helps gas mileage and cruising by advancing the timing when the throttle is partially closed. Many racers leave off the vacuum advance altogether, as they're more concerned with full throttle power. On the other hand, if you are racing with a turbo or supercharger, you must have some sort of mechanism to retard the timing under boost.

The centrifugal advance mechanism works by having a set of counterweights held back by springs. As RPM increases, the weights move outward until they hit a set of stops that limit the total advance. Softer springs will cause the weights to hit their maximum advance at a lower RPM. Finding the right combination of springs can be very much a trial-and-error process. If you can't afford to have the engine dyno tested, your best bet may be to copy the combination used by a proven engine similar to yours and adjust the timing at idle until you get the best performance.

The vacuum advance can also be adjusted, although this may require even more parts swapping. Many original equipment advance units are not adjustable. You may be able to find an adjustable unit that fits your distributor in a junkyard, and if not, there are aftermarket units available. If you are working with forced induction, sometimes you may be able to find a vacuum advance that also retards the ignition under boost.

The alternative to working with springs and advance canisters is to add electronic controls. In some cases, you can simply block off the centrifugal advance, disconnect the vacuum advance (making sure to remove the hose leading to the distributor and cap it at the other end), and install a computerized timing control. Some of these use a series of pushbuttons or a connection to a laptop computer, while others can simply be tuned by turning a few knobs.

Smaller adjustments call for less drastic measures. Some add-on electronics do only one function. For example, the MSD Boost Timing Master simply retards spark timing when a turbocharged or supercharged engine is under boost, using a knob to let the driver control how to adjust the spark. You can even mount the control knob inside the car so you can adjust it depending on what sort of fuel you have in the tank. Other systems have other specific tasks, such as changing the timing when you add nitrous oxide.

CHAPTER **8**

Drivetrain

Putting as Much Power to the Road as Possible

The drivetrain includes everything that carries power from the engine to the wheels. There are two different reasons you might want to install drivetrain mods. The most obvious is that your heavily modified engine might break parts that were not designed to handle your new performance. However, sometimes drivetrain mods can also help you go faster or use less gas. Some upgrades waste less power than the original parts, while others keep your engine turning at the optimum

Once you are getting more power out of the engine, you may need some more parts to get the power to the ground. (*Photo courtesy The Driveshaft Shop.*)

speed to make the most power. A few drivetrain mods, such as performance differentials, can also help reduce wheelspin, putting more power to the ground.

Horsepower by itself usually does not break a drivetrain. The leading killer of drivetrain parts is torque. Adding naturally aspirated mods to the original engine is not as likely to break drivetrain parts as mods such as forced induction or engine swaps. Adding more torque that just spins the tires is also less likely to break parts. Once you manage to get some sticky tires or tune the suspension right, however, the drivetrain suddenly has to deal with the full level of torque. Putting a set of drag slicks on your tire-smoking street car may suddenly cause you to break parts that had been holding up just fine with your street tires.

While toughness is an important concern in a performance drivetrain, light weight is also good to have. Rotating weight slows a car down more than ordinary weight. To see why this is so, imagine a 0–60 test where the car has just reached 60 mph in second gear. On this particular car, the engine is turning at 6,000 RPM at this point, and it has a 12-inch-diameter flywheel. The outside edge of the flywheel will be spinning sideways at 214 mph. This is on top of the fact that the edge is also moving forward at 60 mph. The engine not only has to bring the car up to 60, but it must accelerate the edge of its flywheel to 214 mph. In fact, unless the driver started out in second, the car must bring the edge of the flywheel up to 214 mph twice. Making a part rotate requires extra energy.

Gear Ratios

To understand drivetrains, one must first understand gear ratios. Gear ratios are often expressed as either a mathematical ratio such as 4.11:1 or simply a number like 4.11. If this gear ratio were found inside an axle, its wheels would turn once for every 4.11 turns of the input shaft. The wheels would also have 4.11 times as much torque, but the amount of horsepower going in would be the same as the amount of horsepower coming out (except for friction losses).

Lower gears make the wheels turn more slowly. Since automotive gear ratios are written in an upside-down way, this means that the lower the gearing, the higher the number used to express the gear ratio. An axle with a 5.37:1 ratio is geared "lower" than one with a 3.23:1 ratio. To confuse matters further, some people refer to ratios with larger numbers as "numerically higher." So a "numerically higher" gearing set would also be called a *lower* gear.

The wheels must turn more slowly than the engine because virtually all modern car engines turn much faster than the wheels. If you hooked up an engine directly to the wheels with no gear reduction, the car would need to be going over 100 mph before the engine reached the operating speeds where it works well. Consequently, an engine must be geared down to turn the wheels. Engines also run their best in a narrow RPM range. A car with a single gear ratio might run well from 30 to 50 mph but feel sluggish below 30 and "run out of breath" above 50. This is why transmissions allow you to select several different gear ratios inside them; each speed on the transmission means a different internal gear ratio. A well-set-up car will have its gear ratios chosen so that the engine will spend as much time as possible in the RPM range where it makes the most power.

Manual Transmissions

Many enthusiasts prefer a stick shift. Not only does a manual transmission waste less power than an automatic, but it lets you feel more connected to the car and in control. Their simplicity also means that there is not much the average street enthusiast needs to do to modify the transmission itself. While there are a few times you may want to make internal mods to a transmission to make it more durable or for an all-out race car, complete transmission swaps tend to be more common. Typical reasons to swap transmissions include to make engine swaps simpler, to get a stronger transmission, or to get more speeds. If you have an old three-speed manual from the '60s, upgrading to a modern five or six speed with overdrive can make for much better mileage and more relaxed cruising.

Overdrive gears are the exception to the rule that gears in a drivetrain reduce RPM. An overdrive gear will turn the transmission output shaft faster than the engine RPM. Overdrive gears allow the engine to turn slowly at highway speeds, bringing up the gas mileage.

If you are swapping a front-wheel-drive transmission, also known as a transaxle, there are many pitfalls to watch out for. The transaxle may not put the axle halfshafts at the same distance from the engine as the original. In some cases, the engine oil pan may not clear the transaxle. Even if these parts fit, you will probably need to have a machine shop drill new mounting holes or fabricate spacers to put the parts together, and you may need custom motor mounts or other changes. It is usually best to stick with a transaxle made for your engine. There are a few well-known exceptions. For example, General Motors used very similar mounting designs for the majority of their transaxles, allowing such drastic swaps as putting a Cadillac Northstar V8 into a Fiero while keeping the original transmission.

Rear-wheel-drive transmissions are somewhat easier. Most rear-wheel-drive transmissions have an adapter called a bellhousing between the gearbox itself and the engine. This bellhousing unbolts, allowing you to attach the transmission to another engine by simply installing the correct bellhousing. If the transmission you want, or a very similar one, was ever put behind the engine you have from the factory, you may be able to find a bellhousing in a junkyard. Some aftermarket companies offer bellhousings

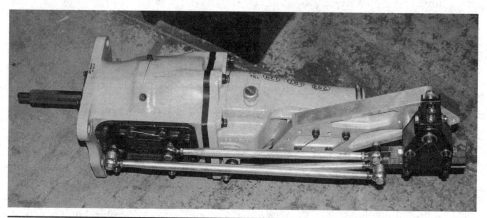

G-Force is one company that offers complete manual transmissions for rear-wheel-drive race cars.

This QuickTime bellhousing adapts a GM transmission behind a Toyota 2JZ-GTE.

to allow installing modern five- and six-speed transmissions in older cars, while companies like Richmond also offer aftermarket five-speed transmissions designed to fit a variety of stock bellhousings.

If you have an automatic and wish you didn't, swapping to a manual transmission may not be as easy as unbolting the original transmission and bolting in a new one. At the very least, you will need to track down all the extra parts used on a manual transmission version of your car—clutch pedal, smaller brake pedal, and all the linkages used for the clutch and shift mechanism. Many newer cars use slightly different engine tuning for different transmissions, or even have the engine control unit (ECU) also run the transmission. You may need an ECU or other engine parts from the manual transmission version of your engine. In the worst case, you may even need to fabricate some of these parts if your car never came with a manual transmission from the factory.

Shifters

If you do not like the way your shifter feels, the aftermarket may have a fix for you. A short-throw shifter is one designed so that you may only have to move the shift knob half as far to go between gears. This can help you shift faster and may improve the way the shifter feels, although a short-throw shifter has to be pushed harder to compensate for not moving as far. A good quality shifter will also improve feel by being built to tighter tolerances so there is less "slop" and vagueness in the way it feels. There is no shortage of shift knobs to top off your shifter, from leather to carbon fiber to metal to models with neon lights. Keep in mind that installing a taller shift knob may make shift throws longer, and a metal shift knob may become uncomfortably hot on summer days.

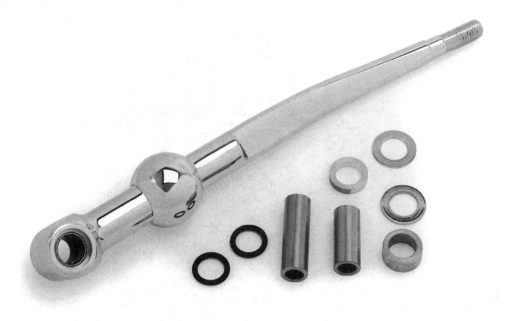

A short-throw shift kit. (*Photo courtesy Edelbrock.*)

Flywheels

The flywheel stores energy to keep the RPM from dropping too much when you release the clutch. Unfortunately, a flywheel's heavy weight can also slow you down. Accelerating the flywheel by a couple thousand RPM for each gear requires power. Replacing a heavy steel flywheel with a lightweight aluminum one, or sometimes simply a steel flywheel with less material, can make the car accelerate like it is much lighter. The downside is that a light flywheel may take more careful work with the clutch to get the car moving.

Keep safety in mind when looking into flywheel mods. Not only can the flywheel spin at speeds exceeding 200 mph, but they are covered on the outside with over a hundred gear teeth where the starter connects. If one of these breaks, those teeth could easily come through the floorboards and attack your feet. Trying to get a lightweight flywheel on the cheap by machining off some of the material can be very risky. Good quality flywheels will usually have SFI certification (SFI is the main organization in the United States that certifies racing parts are safe) or its foreign equivalents. If you are especially concerned about flywheel safety, or if you race your car, you can buy a reinforced scatter shield that installs in your transmission and can contain an exploding flywheel. If you are shopping for a scatter shield, be sure to buy one with SFI certification.

Clutches

If your clutch starts to smell burnt after a hard launch at the dragstrip or it starts to slip even when your foot is off the clutch pedal, it's time to upgrade to a better clutch. The most common performance clutches copy the basic design of an original equipment

A conventional single-disc, full-face clutch. Note the sprung hub. (*Photo courtesy Spec Clutch.*)

clutch, which pushes a single disc of material similar to brake pads against the flywheel using a device called a pressure plate. A clutch with a single disc of friction material is also called a full-face clutch. Holding power can be increased by using grabbier clutch materials like Kevlar or carbon, using stiffer springs to push the clutch material against the flywheel, or just plain making the clutch bigger. Keep in mind that high-performance racing clutches may require a bit more force on the clutch pedal, and may be harder to use smoothly.

Puck-style clutches have several sections covered with friction material, usually four or six. Puck clutches, particularly with smaller numbers of pucks, tend to jump rather abruptly from just barely grabbing to fully engaged. This is a popular design for racing, but not the easiest type of clutch to use in a daily driver.

Most production clutches have a sprung hub, with springs that absorb vibrations and chatter. Some racing clutches have a solid hub, either to fit more pucks or to reduce weight, or because this is one less part to break behind a torquey motor. Solid hubs result in very harsh, chattery clutch engagement. While you can still be somewhat smooth with a four-puck clutch and a sprung hub on the street, a solid hub is one clutch feature you should save for pure race cars. The one exception is on a car with a dual-mass flywheel, which relocates the spring assembly to the flywheel instead of the clutch.

A six-puck clutch with an unsprung hub. *(Photo courtesy Spec Clutch.)*

A dual-disc clutch assembly. (Photo courtesy Spec Clutch.)

If one clutch is not strong enough, use more. This is the idea behind multiple-disc clutches. These combine a lightweight flywheel with a stack of two or more clutches. These cost a good bit more than a normal design. If your car regularly eats high-performance clutches, however, this can be money well spent.

Automatic Transmissions

Automatic transmissions have several strikes against them in the performance world. Not only do they slip and waste power, but if one shifts at the wrong time in a race, it can upset a car's cornering balance. Consequently, you may be surprised to hear that there are more mods available for many automatic transmissions than for manual ones. Automatics are popular with drag racers because they can make shifts much faster than a manual transmission and have several beneficial effects from the torque converter, the fluid coupling that connects the engine to the transmission.

While there is a lot that can be done with the torque converter, most mods to an automatic transmission are either to make it more durable or to make it shift faster. Installing durability mods often takes an expert, but there are relatively simple do-it-yourself packages that let you install a shift kit for firmer shifts. Shift kits can make a transmission shift more harshly, but interestingly enough, this actually puts less wear on the transmission. An automatic transmission uses a complicated system of small brakes and clutches inside it to select gears, and slow shifting lets these slip for a longer time, putting more wear on the transmission.

Some racers actually take the automatic part out of the automatic transmission entirely with a mod called a manual valve body, which makes the transmission only

There are a lot of parts available for automatic transmissions. (*Photo courtesy Comp Performance Group/TCI.*)

This shift kit replaces the valve body. (*Photo courtesy Comp Performance Group/TCI.*)

shift in response to the shifter. Some manual valve bodies use a "reverse pattern" that flips the order of the numbers on the shift lever, so first is where D used to be and top gear is where the 1 is on a normal shifter. If you get a reverse pattern valve body, you can also buy aftermarket shifters labeled to match this pattern. In case you were wondering why somebody would bother flipping the shift order upside down, this is to make it harder to accidentally shift into reverse during a race. Many aftermarket shifters also include a reverse lockout, requiring you to push a button to put the car into reverse for additional safety.

One durability mod that is quite easy to install is a transmission cooler. Most cars already have a small transmission cooler located in the radiator that also helps the transmission warm up on cold winter days. In most cases, the best place to install an aftermarket transmission cooler is in the lines to the original cooler on the upstream side, so the fluid goes through the cooler first and then the radiator.

The transbrake is a mod that sometimes appears in beefed-up drag racing transmissions. A transbrake allows the driver to put the transmission into first gear and reverse at the same time. Naturally, the car cannot go anywhere in this condition. This lets the driver hold the engine at full throttle and then suddenly release it for hard dragstrip launches. Transbrakes are usually part of an aftermarket valve body.

Swapping an automatic transmission can be somewhat more complicated than swapping manual gearboxes. Most automatic transmissions have the bellhousing and transmission case cast as one unit, instead of a detachable and easily swapped separate bellhousing. In some cases, an aftermarket transmission shop may offer transmissions

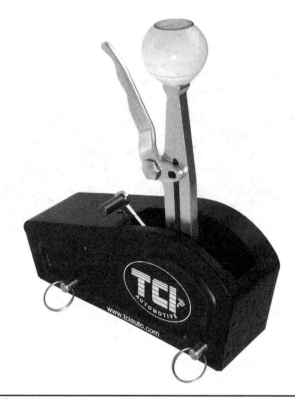

An aftermarket shifter with reverse lockout. (*Photo courtesy Comp Performance Group/TCI.*)

that have been machined to fit different engines. Sometimes it is possible to install the internal parts from a stronger transmission into your original casing. In either case, installing a transmission that will not easily bolt in is best left to experts. One nonexpert

TCI offers controls for many popular automatic transmissions. (*Photo courtesy Comp Performance Group/TCI.*)

option you may be able to find is an adapter plate that goes between the engine and the transmission. If you can find one of these for your engine and the transmission you want, the swap can become an easy bolt-in.

Some modern automatic transmissions have computer controls. If you wish to install one behind an older engine without the right electronic controls, you will need to buy an aftermarket control box for your transmission.

The Mysterious Torque Converter

For a device with only three moving parts, a torque converter can be complicated to understand. The parts inside a basic torque converter are the impeller, the stator, and the turbine, which sit in a housing full of transmission fluid. The impeller is shaped somewhat like a hollow half-bagel filled with fins. It connects to the engine and gets the fluid moving. The turbine is a mirror image of the impeller and transmits the motion of the fluid to the output shaft. The stator is a finned disc that fits between the two. A one-way clutch connects the stator to the transmission housing.

At steady cruising speeds, the turbine spins at almost the same speed as the impeller. The stator freewheels at a speed somewhere in between. When the car is stopped and the engine idles, the fluid slows down to the point that the torque converter almost behaves as if the two ends are completely disconnected from each other, and the stator stops moving. This "declutching" is why torque converters work well with automatic transmissions; they can disconnect at idle without any help from the driver.

Things become more complicated when the car takes off from a stop. When the impeller suddenly pushes the fluid harder, it rams the stator back against its one-way

The right torque converter is key to making an automatic transmission work in a drag race. (*Photo courtesy Comp Performance Group/TCI.*)

Inside a torque converter. (*Photo courtesy Comp Performance Group/TCI.*)

clutch. The fins on the stator redirect the fluid so that it leaves the stator at a different angle, pushing the turbine forward with a torque equal to the total of the incoming engine torque plus the torque the stator puts on its clutch. The turbine now turns with more torque at a slower RPM, in much the same way that gears can reduce RPM while raising torque. The more the torque converter slips, the greater the torque multiplication. The fluid acts more smoothly than a mechanical system, so the stress is not as high as when a clutch suddenly engages. Because the shock is less, some cars with automatic transmissions use weaker rear axles than their stick shift counterparts.

There are several performance considerations when selecting a torque converter. In many cases, you may want to find a torque converter company with a good reputation and talk with them to have them recommend a torque converter especially for your car. They may have a torque converter off the shelf that will suit your car, but some applications may call for a custom torque converter that can cost $1,000 or more. If you are picking a torque converter out of a catalog, you will probably only be told three things besides the brand name and price: its stall speed, diameter, and whether the manufacturer has done anything to make it stronger.

Torque converter stall is somewhat complicated. To understand how torque converter stall speed works, imagine that you have a model torque converter sitting on a table. This imaginary model comes mounted on a display stand complete with a crank that lets you turn its housing and a pair of gauges. One of the gauges shows the RPM at which you are turning the crank and the other shows the amount of torque you are applying to the crank. The output shaft on this torque converter is clamped to the display stand so it cannot turn.

Now imagine that you start to turn the crank and are spinning it at 10 RPM. The input gauge shows that you are putting 5 lb-ft of torque on the crank. Try bringing the RPM up to 20. This takes some more effort. The gauge will show that this takes 20 lb-ft of torque. Bringing this torque converter up to 30 RPM would take 45 lb-ft of torque, at which point you would probably have to hold onto the stand carefully to prevent it from tipping over. This particular torque converter happens to be sized to make it suitable for curious tinkerers to play with instead of for use in a car, but it shows a general principle of how torque converters behave. If you hold the output shaft still while the turbine is moving, a condition known as stall, it takes roughly four times as much torque to double the RPM.

The stall speed is the fastest a particular engine can turn a particular torque converter when the output shaft is held still. Stall speed is not a property of the torque converter itself, but of the engine and torque converter combination. A torque converter company can predict this speed by putting a torque converter through a test similar to the one described for the model torque converter. They can then produce a graph of torque versus RPM to show a stall curve for the torque converter. If you plot this graph on the same scale as the engine's torque curve, the stall speed is the point where the two curves intersect. Manufacturers will advertise a range of stall speeds based on the sort of engine they expect it to be attached to. If you could take a torque converter meant for use behind a big block V8 and meant to stall at 2,000 RPM and adapt it to fit a Honda Civic, this converter might have an unacceptably low stall speed below 1,000 RPM.

The preceding example is known as true stall, the highest speed an engine can maintain with the output shaft held still. Another rating is flash stall speed, which is the fastest speed an engine can hit when it is revved from idle. The flash stall speed is a little higher than true stall. In some cases, the brakes may not be strong enough to hold back an engine when it reaches true stall speed, so you may have to settle for leaving the line at lower RPM unless you can install a transbrake.

For drag racing, you will want to select a torque converter with a stall speed only slightly below the engine's peak torque, allowing you to rev your engine up to where it will perform strongly before your car is moving. The downside is that, while a torque converter can drive a car at well below its stall speed, it will get less gas mileage, generate much more heat, and generally not behave like a good street car. A "loose" (high stall RPM) converter could cause your RPM to jump surprisingly fast if you tap the gas pedal. Not a good thing for stop-and-go traffic. Selecting the right converter for a heavily modified engine is often a compromise between dragstrip performance and good street behavior. Since most naturally aspirated mods raise the point where an engine makes maximum torque, serious internal mods often call for a higher converter stall speed for the best performance. Worse, if the mods reduce your low RPM torque, they may pull down the converter's current stall speed.

You might expect a large-diameter torque converter to be able to withstand more torque. This would be true, except that their internal parts are often made of the same thin sheet metal as smaller-diameter converters. Larger torque converters do feature more torque multiplication virtually all the time. The downside is that behind a very high-powered engine, this extra torque multiplication can break the fins inside it. This can lead to the paradoxical result that sometimes you may need a smaller-diameter torque converter to handle a more powerful engine. Smaller-diameter converters also work well at high stall speeds, making them well suited to high-revving, high-powered engines. Larger torque converters work well in heavier cars and mildly modified V8 engines.

Torque converter builders have several tricks they can use to strengthen a converter. The fins on the moving parts are normally tack-welded on, a process that leaves them held in place by only a thin spot of metal. Furnace brazing secures the fins all around their edges with solder for extra strength. Antiballooning plates or similar reinforcement help the housing withstand high pressures. Heat treatment can strengthen the steel. Another common tweak is installing needle or Torrington bearings to reduce friction and survive stronger loads.

Selecting a torque converter is quite complicated. Even if you are ordering an off-the-shelf torque converter out of a catalog, you may wish to call the manufacturer directly before ordering and ask which converter they recommend. The manufacturers will be happy to offer their suggestions as to what would be a good match. They will want to know about your engine, the car weight, the transmission and gearing, and how often you race versus driving on the street (not being honest about that last point might come back to haunt you). Common engine and car weight combinations can often use an off-the-shelf converter. If you upgrade to a high-stall converter, be sure to install a transmission cooler to deal with the extra heat.

Many times you will see a torque converter described as "rebuilt" or "remanufactured from select cores." Very few aftermarket converters are entirely new. Most of them are based around the "core" of a used torque converter with some of the internal parts replaced.

A lockup torque converter combines a torque converter with a clutch. Once the car reaches cruising speeds, the clutch engages to prevent the torque converter from slipping and to boost mileage. The lockup feature is most important with an overdrive transmission, where the combination of low RPM and a fair amount of torque could result in a lot of heat build-up when cruising in overdrive. There is some controversy about whether there is any performance gain from using a lockup torque converter; some racers remove the lockup feature entirely, while others retain it.

Driveshafts and U-joints

U-joints and their cousins the CV joints show up in several places in the drivetrain. Both of them are designed to put a joint in a rotating shaft so it can bend. The CV joint, or constant velocity joint, can work at larger angles than a U-joint, making it a good choice for getting power to the wheels if you do not have a solid axle. U-joints are simple and inexpensive, but only work well at small angles before vibration becomes a problem. These are usually found on the driveshaft running from the transmission to the rear axle or differential, which does not move nearly as much as the shafts leading to the wheels themselves. If these joints are not strong enough, fixing this is simply a matter of sourcing a stronger set of joints and having them installed on your drivetrain. Check with fellow enthusiasts to find out how much power you need to break your drivetrain's joints and where you can find stronger parts.

A driveshaft can be a very simple part—just a tube with a U-joint at each end. Sometimes swapping engines, transmissions, or axles can change the distance between the transmission and the differential. While you may be able to use the driveshaft from the donor car, you might need to have one custom made. This is quite cheap compared to most custom work. A driveshaft maker only needs to cut a piece of tubing to length, weld on the appropriate U-joints at each end, and balance the assembly. Most cities have shops that can make basic steel driveshafts. Adding strength to the tube is simply a matter of using a wider tube, thicker tubing walls, or stronger steel.

While steel driveshafts are quite functional, racing driveshaft makers also sell lightweight driveshafts made from aluminum or carbon fiber. These not only reduce weight, but if you have a solid axle rear suspension, can improve the ride quality a little. Ordering one is simply a matter of telling the driveshaft shop how long your driveshaft is, what sort of joints you need at each end, and how much power and torque you need the shaft to survive.

U-joint on the end of a carbon fiber driveshaft. (*Photo courtesy The Driveshaft Shop.*)

One-piece driveshafts can be installed very quickly. Typically, there are only four bolts that must be removed at the back end to partly disassemble the U-joint, at which point the front will slide out of the transmission. When taking apart a U-joint, be sure to put the caps back on in the exact same location they were originally. Installing a cap backwards or switching the two caps can make the edges of the hole not line up correctly. The cheaper sheet metal straps do not need this amount of precision, but also can't transmit as much torque.

Other driveshafts are more complicated. Some of them have U-joints in the middle, and possibly other features like splined, extending sections and support bearings along their length. The main reason for splitting up a driveshaft is that long shafts can vibrate or even buckle if spun too fast.

A driveshaft safety loop is an essential safety item on rear-wheel-drive or all-wheel-drive race cars. It's also a good idea for any high-powered street car. This is a steel loop that completely encircles the driveshaft and mounts to the subframe a few inches behind the front U-joint. These are typically made of ¼-inch-thick steel straps, 2 inches wide, and need at least 3 inches of clearance around the driveshaft.

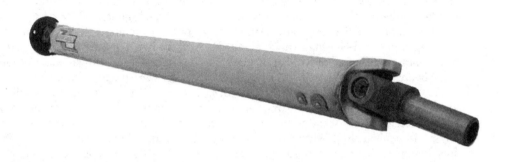

One-piece driveshaft. (*Photo courtesy The Driveshaft Shop.*)

A two-piece driveshaft with center support bearing. (*Photo courtesy The Driveshaft Shop.*)

Driveshaft safety loops are to prevent a broken front U-joint from turning into a full-blown catastrophe. If the front U-joint breaks, the driveshaft is only attached at the rear of the car. Unless there is a loop to catch it, a loose driveshaft can easily dig into the pavement and damage the rear suspension. A driveshaft safety loop is cheap insurance, and required for many racing events.

No such loop is usually needed at the rear. Having the driveshaft dangle by its front U-joint is a much less threatening situation unless you are driving at high speed in reverse. Ordinarily, the worst that may happen if the rear U-joint fails and the driveshaft drops to the pavement is that you might need a new driveshaft and need to beat a few dents out of your floor pan. The driveshaft hitting the pavement is much less likely to send you off in a sudden, unexpected direction.

Gearing

After the transmission, the drivetrain uses another set of gears to reduce the speed to one suitable for driving the wheels. The ratio of these gears is known as the final drive ratio. On many rear-wheel-drive cars, it is possible to buy gears with different ratios. The larger the number used to express the gear ratio, the higher an RPM your engine will turn at any particular speed. Gears with larger numbers are better suited to high-winding engines and letting a car accelerate hard off the line, but overdoing things may mean your car runs out of RPM at the end of a drag race. Gears with smaller numbers work better for relaxed cruising if you do not have an overdrive transmission.

If you have changed your gear ratio or tire diameter, your speedometer will no longer read correctly. On older cars, correcting this is usually a matter of replacing the speedometer drive gear in the transmission. Most companies that sell these gears will have a convenient chart letting you look up your gear ratio and tire size to pick the right speedometer drive gear. With some newer cars, correcting the speedometer may require reprogramming the ECU.

A final drive gear in a 9-inch Ford axle. (*Photo courtesy The Driveshaft Shop.*)

Differentials

When a car goes around a corner, the wheels on the outside turn faster than the wheels on the inside. The drivetrain needs to allow for these different speeds. The simplest way to do this is through a set of gears called an open differential. This uses a pair of gears called side gears that face each other, with each side gear driving one wheel. The side gears are mounted in a differential carrier that is turned by the transmission. The differential carrier has one or more shafts running from its center between the side gears, perpendicular to the main axle halfshafts. These shafts spin with the carrier. Each shaft has a pinion gear that is free to rotate about the moving shaft. Each pinion gear meshes with both side gears. When the car moves in a straight line, the side gears turn at the same speed as the carrier, and the pinion gears will not rotate. If you could attach a miniature video camera to your differential carrier and point it at the gears, the camera would show the gears standing still. When the car turns a corner, the gear on the outside moves faster than the carrier, the gear on the inside moves slower, and the pinions turn. The view from the imaginary differential-cam would show the side gear on the outside slowly drifting forward and the side gear on the inside drifting slowly backward.

Contrary to popular belief, an open differential is not one-wheel drive at all. The open differential will always send an equal amount of torque to both wheels. By nature, it cannot deliver more torque to one wheel than the other. The problem is that if one wheel loses traction and starts to spin, the differential cannot send much of any torque to that wheel. Consequently, it cannot send any torque to the wheel that still has traction, either.

Spinning the tires like this is not acceptable for high-performance driving. Sometimes this can be solved using stickier tires or suspension tuning. Sometimes it cannot.

The opposite end of the spectrum is to remove the differential or prevent it from turning, making what is known as a locked differential. The cheapest way to do this is to simply weld the gears in the differential to each other. A stronger option is to replace the entire differential with a solid piece of metal called a spool. This works just fine for drag racing. It's not too bad for driving on dirt, either. Unfortunately, if you take a car with a locked differential around a paved corner, the tires can only accommodate this by scraping off their tread. This generally isn't good for handling on pavement unless you only take gentle, sweeping turns.

Most high-performance differentials are designed to split the difference between an open differential and a spool. These differentials, by one method or another, transfer a part of the torque between the axle shafts. Some of them can act almost exactly like an open differential when no power is applied, but partially lock up when you get on the gas. These are known as one-way differentials because they only work when transferring power one way—that is, forward. A two-way differential, by contrast, will work in the same way whether you are accelerating, braking, or taking a corner with your foot off the throttle. A differential that transfers some torque at all times, but transfers more when given power, is known by the somewhat confusing term of a 1.5-way differential.

Performance differentials are not cheap. Some of the more common American solid axle designs have differentials available for $500, but depending on your car and the type of differential, you may have to spend over $1,500 for parts alone. Installation costs can also run in the hundreds of dollars if you don't handle the work yourself.

Performance differentials can work on three different principles. Limited slip differentials use various means to create friction between the two wheels if one starts spinning. This design includes the clutch type, the viscous differential, and the cam and pawl. Torque biasing differentials use a mechanical system that can automatically send the most torque to the wheel with the most available traction, even before the differential starts to slip. The only common design that works this way is the helical differential. The third type, locking differentials, completely lock the two sides together under power.

The most common sort of performance differential is the clutch type. This uses a set of clutches attached to the side gears to create friction between the side gears and the differential carrier if the side gears turn at a different speed from the carrier. The simplest use cone-type clutches that cannot be easily adjusted or repaired. The sort normally called a clutch-type differential uses several layers of flat-plate clutches.

There are two ways to apply pressure to the clutches to make them work: preloaded springs and ramps. Preloaded springs simply use a spring to force the clutches together. This keeps the differential partially locked, regardless of what the car is doing. Some preload on the springs can work on rear-wheel-drive cars, but preloading the springs should be kept to a minimum for most front-wheel-drive cars. If the differential uses ramps, the pinion shafts are not entirely still, but move slightly when the engine applies power. The pinion shafts move against a mounting system that forces the clutches outward, tightening up the differential under power and loosening it when you lift off the throttle.

Clutch-type differentials are available for a wide variety of cars and can work well for most sorts of racing. Serious racers often prefer plate clutches over cone clutches, but cone clutches can work well for street cars too. The biggest downside of a clutch-type differential is that it requires periodic maintenance and proper break-in to work its best.

A few cars use "active differentials." Most of them are clutch-based, but use solenoids or hydraulics combined with computer controls to change the amount of clamping force on the clutches. The only way to tune one of these is to reprogram the computer that controls it.

Occasionally you may see clutch-type differentials made by modifying stock units to add clutch plates. These are often very cheap compared to other units, but have many drawbacks compared to a purpose-built clutch differential. The largest disadvantages are that they often require very large amounts of preload and have a reputation for wearing out quickly. From a performance standpoint, they can still be better than an open differential, but they often require even more frequent maintenance than other clutch-type designs.

Clutches do not need to be engaged by engine torque or spring pressure, however. One design, called a speed-sensitive, progressive-locking differential, uses an oil pump driven by the difference in speed between the two side gears. This pump creates oil pressure that squeezes the clutches together. These operate very smoothly and offer a wide range of adjustment, but are only available for a few cars at the moment. Confusingly, GKN calls their version of the speed-sensing, progressive-locking differential the Visco-Lok, even though it is not a viscous differential at all.

Viscous differentials fall into two categories. The one most commonly seen on the street combines an open differential with a fluid coupling, similar to a torque converter with no stator, connecting the two side shafts. The greater the difference in speed between the wheels, the more torque the coupling transmits. Viscous differentials like this are often found on various Nissans. This sort of differential can be useful for getting a car out of the snow, but it reacts much more slowly than other designs. Most enthusiasts prefer to run a more aggressive sort of differential.

The second sort of viscous differential eliminates the clutches and gears altogether. Each shaft is attached to a series of discs. The carrier also has a set of discs that run in between the discs on each side shaft. The differential is filled with a thick, sticky, silicone-based liquid that transmits the torque from the carrier to the shafts. This design has very little about it that wears out, and it reacts instantly to any difference in speed. It is normally used for road race cars. On the street, the way both shafts slip can throw off the accuracy of your speedometer.

The cam and pawl differential is another gearless arrangement. Each side axle is connected to a cam—in this case, the cam looks like a metal cylinder with a zigzag circle cut into its end. The engine turns a moving cylinder that fits in between these two cams. This cylinder, sometimes called a cage, contains several square pieces of metal called pawls. These pawls are often nicknamed "chiclets" because they are about the same size and shape as the little bits of chewing gum. Under acceleration, the pawls wedge themselves firmly into the cams to transmit power to each axle. If one wheel starts spinning faster, the pawls wiggle in such a way as to let the wheel spin faster but not give it more power, keeping a wheel that has lost traction from spinning madly. Cam and pawl differentials work in a very forgiving and predictable way, making them a great choice for someone getting into autocross or road racing with very little experience. The downside is that they wear out much faster than most other differential designs.

The helical differential is the only common torque-biasing differential on the market. These differentials replace the standard arrangement of pinions and side gears with more complicated gears that only operate efficiently in one direction. The result is that the helical differential will automatically transfer torque to the wheel with the most traction. Helical differentials are also maintenance free, with some companies offering a lifetime warranty even for race cars. This differential works very well for autocross, road racing, and street use. It also works reasonably well for drag racing. However, it can have somewhat treacherous effects at high speeds on dirt roads because it can suddenly put more torque toward the side with more traction and actually cause the car to steer sideways. These differentials also behave like an open differential if one wheel is off the ground or has absolutely no traction, but if this happens, stabbing the brakes while giving the engine a little gas can pull you out. The best known helical differentials are made by Quaife and Torsen.

The Detroit Locker is the most common locking differential on the market. Each side is connected to the carrier using a clutch instead of gears. If one wheel starts to move faster than the other, the Detroit Locker simply unlocks its clutch and sends all of the power to the other wheel. Usually, it will completely lock under power. This design is brutal, noisy, and usually not the best choice for road racing unless you are running in NASCAR where the rules require it. However, it works very well for drag racing or dual-purpose street/drag cars.

The term "locking differential" is sometimes also applied to a differential that normally operates as an open differential but can completely clamp together like a spool at the flip of a switch. These usually show up on off-road trucks, but could also be used on a street/drag car.

While a two-wheel-drive car uses just one differential, an all-wheel-drive car will need three. The typical all-wheel-drive car features a center differential that attaches directly to the transmission, mounted in a part called the transfer case. This differential sends power to the differentials located at the front and rear of the car. The three differentials on an all-wheel-drive car are not necessarily all the same design. It is possible to run a cam and pawl front differential, an active clutch center differential, and a viscous rear differential if the designer wants.

Some trucks and SUVs omit the center differential, instead connecting the rear wheels permanently to the engine and using a clutch or other device to connect the front wheels when shifted into four-wheel drive. This system is excellent off road, but this drivetrain can only safely operate in two-wheel-drive mode on pavement.

Axles and Halfshafts

Halfshafts deliver the power from the differential to the wheels. On a front-wheel-drive car or a car with rear-wheel drive and independent rear suspension, the shafts are quite similar to driveshafts, solid steel rods ending in CV joints. Turbocharged front-wheel-drive cars with sticky tires are especially prone to breaking halfshafts. The cure is, of course, to strengthen the shaft by making new ones with a larger diameter or from a stronger type of steel such as 4130 or 4340 (number codes for certain steel alloys). Larger shafts will almost always require larger CV joints that fit over the oversized shaft, even if the CV joints are not the weakest link.

High-powered front-wheel-drive cars sometimes have handling problems if the halfshafts are different lengths. Letting the halfshafts run at different angles can cause an effect called torque steer, where the torque delivered by the shafts causes the wheels to toe inward at different amounts on each side under full power. There are some cases where it is possible to convert a car with unequal-length halfshafts to equal-length units by adding an intermediate shaft on the longer side. The intermediate shaft is a straight

Halfshafts for an independent rear suspension. (*Photo courtesy The Driveshaft Shop.*)

rod extending from the differential that is not free to pivot, moving the mounting point for the halfshaft outward.

Halfshafts inside a solid axle are a little different. These typically are entirely one piece with the wheel end flared out into a disc and the other end having a set of splines. Splines are a sort of ridge or groove running along the length of a shaft that fit into a matching set of splines in the differential. The strongest halfshafts use a small radius on the point where they flare out and keep the exact same diameter from the inside of the flange to the splined area. A shaft is only as strong as its thinnest section, and keeping it a constant diameter helps the shaft absorb the sudden shock of a hard launch.

These shafts are typically rated by the spline count. The number of splines per inch is standardized, so all axles with the same number of splines have the same diameter. More splines make for a stronger axle, but you will also need a differential with the correct spline count. Companies like Strange and Moser offer custom axle shafts built in any length and spline count a gearhead could possibly need.

Solid axles need something to keep the halfshafts in the axle housing. Some of the less expensive designs use C-clips located inside the differential. The problem with C-clips is that if the axle shaft breaks at some point in the middle, the outside part of the shaft is going to come right out. Other cars use bearings at the wheel that hold the axle in, making it much less likely that a broken axle will leave the car. Axles with C-clips can be converted to this sort of axle with a C-clip eliminator kit. Some trucks use a third design called a full floating axle where the bearings hold the brake and drum assembly to the end of the axle tube, with the axle bolted to the hub. This design lets the axle only transmit torque instead of bearing the car's weight or carrying side loads. While this is a very strong design, it's also heavy and expensive.

Sometimes the best way to get a stronger rear axle is to swap an entire axle assembly from a donor car. Pick the right axle, and you can get stronger shafts, larger gears, and more possibilities for differential and brake upgrades. Axles are remarkably interchangeable between rear-wheel-drive American cars if they use a leaf spring suspension. The most important factors to check are that the axle is about the same width and has the same distance between spring perches as what your current axle has. If this is the case and the shock mounts are about the same, you can probably bolt it directly into your car. It may require a special adapter U-joint or some driveshaft changes, but it is possible to make such unusual swaps as putting a Ford Maverick rear axle in a Plymouth Duster without major fabrication.

With major fabrication, there are even more possibilities. Axle shops can make an astonishing variety of changes to axle housings, from narrowing the axle to welding on different spring perches or mounts for coil spring suspensions. Some axles are even put together without using any parts that ever left the original factory, and add lightweight parts like aluminum housings. If you want something truly off the wall like a 9-inch Ford axle that fits your Datsun B210, there are several racing shops that can build one for you. The 9-inch Ford axle is probably the most popular design for swapping because it has more aftermarket parts available, but other popular choices include the Dana 60 (perhaps one of the toughest axles ever originally in passenger cars), Chrysler 8¾ inch, and GM 12-bolt. These four axles offer wide choices for differentials, gear ratios, brakes, and halfshafts.

Engine Mounts

Most original engine mounts use rubber cushions to absorb engine vibrations. This makes for a comfortable driving experience, but the softness of the rubber can be a problem for performance driving. Original equipment mounts can bend enough under acceleration for the engine to rock side to side, and adding more power can make them bend even more. I have driven one car where the original engine mounts were so soft

Extra-stiff engine mounts.

that the shift knob would move nearly an inch just by going from full throttle to letting off the gas, and that was with a stock engine.

Preventing the engine from moving too much can help the car react faster and create more predictable handling. On front-wheel-drive cars, reducing engine movement can also improve traction by reducing wheel hop. The largest trade-off is that the car will have more vibration. There are a number of ways to get a stiffer set of engine mounts.

Many modern engine mounts have openings in the rubber blocks. The cheapest way to stiffen the engine mounts is to fill these holes with liquid polyurethane or even polyurethane caulk. In some cases, it is possible to use plastic or metal inserts instead.

For those who want a more professional solution, the aftermarket offers motor mounts in many different materials. Polyurethane is often a good choice for street use. Delrin or metal motor mounts are usually best used for racing. These ultra-stiff mounts transmit enough vibration that they can cause cracks in the area where the mounts bolt to the chassis, so they may require reinforcement or periodic checking for damage on some cars. Another concern is that all the mounts should be of the same stiffness, including the transmission mount (although a transmission mount that is less stiff than the other mounts is less of an issue than a polyurethane transmission mount with rubber engine mounts). If one mount is replaced with metal or polyurethane while the rest are left as stock rubber, the stiffer mount will take almost all the stress of the engine vibration and will be much more likely to crack.

Another sort of rigid engine mount is a motor plate, sometimes known as elephant ears. This design replaces the regular engine mounts with a solid metal plate that attaches to the front of the engine using bolt holes that are already there, such as the holes where the alternator and water pump attach. This normally only works for rear-wheel-drive cars. The advantage of a motor plate is that it frees up more room in the engine compartment for headers, turbo installations, or other exhaust work. Motor plates sometimes relocate the water pump or other parts, so some of them require new pulleys.

If you want to use stiffer engine mounts, it is often worthwhile to invest in having a machine shop take extra care in balancing your engine. Not only will this make driving your car less harsh, but it will put less stress on your engine mounts.

Some cars use a bobble strut, basically a shock absorber designed to limit engine motion. Often, the original bobble strut is too soft for hard driving. Replacing the bobble strut with a stiffer design, or adding a bobble strut to a car that never used one, is often a good way to mount the engine more rigidly. Some rear-wheel-drive cars use a chain or torque strap connected to the left side of the engine for similar purposes.

CHAPTER 9

Suspension

Tuning It to Best Fit
Your Driving Style

Making a car handle well can be somewhat more complicated than improving horsepower. With horsepower, the problem is usually, "Not enough power," and the solution is, "Add more power." Handling is more complicated. There is only one way for a car to go slow, but several ways a car may handle badly. To improve handling, you must know both what is wrong with the car's handling and what changes can correct it.

Selecting the right combination of suspension parts is critical to making a car handle well. (*Photo courtesy Edelbrock*.)

Understeer and Oversteer

These are the two most commonly discussed problems when it comes to handling. Both of them result in the car not heading in the direction the steering wheel should have pointed it. A car that oversteers will rotate more than expected and feel as if the rear tires have started to slide sideways. The tires are not actually sliding, but instead are drifting. Drifting occurs when the force of cornering causes the tire to bend sideways, and the direction in which the car is traveling becomes slightly different from the direction in which the tires are actually pointed. The angle between the direction in which the car is going and the direction in which the tire is pointed is known as the *slip angle*.

When a car oversteers, the rear tires drift more than the front tires. Even though the back wheels are not skidding, the driver may feel as though the rear of the car is trying to whip around and pass the front. In some cases, you may even have to point the steering wheel in the opposite direction of the turn to keep the car going around the corner. Catching a car that starts to oversteer is often tricky, since hitting the brakes—or sometimes even lifting off the throttle—can make the rear tires let go entirely. Race car drivers often describe a car that oversteers as "loose." Street drivers typically describe a car that does in this in normal driving using less socially acceptable language.

Understeer happens when the front tires drift more than the rear tires. A car that understeers will feel as though it is not responding enough to the steering, requiring the driver to slow down and turn the wheel more. Racing terms to describe understeer include "pushing" and "plowing."

Even if a car is understeering, the rear tires may still have a substantial slip angle if the understeer is not too severe. This means that a car can "hang the tail out" even if it is not oversteering at all.

Most cars will understeer in some situations and oversteer in others. Which one the car will do depends on its speed and whether the car is accelerating or braking. Since recovering from understeer is often more intuitive and less likely to send the car into an uncontrollable spin if the driver reacts too late, most production cars leave the factory set up to understeer except in extreme cases. To build a good handling car, you will want to balance the car on the edge between these two, a condition called *neutral steer*.

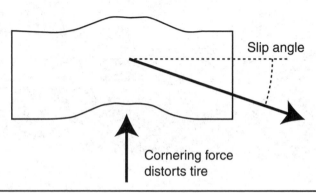

A tire distorts under cornering forces, causing the car to travel in a different direction from the one in which the tire is pointed.

A car with neutral steering can keep a large slip angle on all four tires when pushed to its limits, a condition known as four-wheel drift.

Underdamping

Underdamping is a problem that occurs when the shock absorbers or struts are not matched to the springs. An underdamped car will show different sorts of misbehavior depending on how badly the parts are mismatched. In extreme cases, a car may bounce violently when driven over rough roads. Less underdamped cars may feel as though they are floating rather than solidly connected to the ground, or have the suspension wobble two or three times when the car hits a bump. The car may pitch violently forward when you apply the brakes, rock back and forth when you shift gears, or lean more quickly when going into a corner. Other symptoms can include handling that simply feels vague and sloppy, or bottoming out on speed bumps that the car looks tall enough to clear. Underdamping is a common problem in cars that have worn-out suspensions or that have been improperly lowered. These sorts of problems will typically show up when driving over bumpy roads, or when the car is just entering or exiting a corner.

Bump Steer

Bump steer is a suspension issue that causes the tires to change the direction they are pointed in when the car hits a bump. A car with this problem will behave fine on smooth pavement, but acts like it "has a mind of its own" if you drive on bumpy or broken pavement.

Just Plain Worn Out

Suspension parts wear out over time. Metal scraping together can wear down, and old rubber can fall apart. An excessively worn suspension may make banging noises on bumpy roads, and often will wander over the road if you try keeping the steering wheel pointed in one direction. If something's worn out or broken in your suspension, you'll want to fix this before you try any other upgrades.

Other Handling Issues

Most remaining issues are easier to recognize. Some cars may feel twitchy when going in a straight line, or have the opposite problem, responding too slowly when you want them to turn. When cornering hard, many cars lean and roll, or their tires simply let go too early. A suspension may bottom out, either by going over bumps or while taking a corner. If the suspension bottoms out on a corner, the end where the suspension bottomed out can immediately lose traction. And, of course, there is always the question of how comfortable you want your car to be.

Suspension Tuning Basics

Many handling issues do not result from a problem with one component, unless that part is broken. Handling is determined by the way the parts of the suspension interact.

There are often several different ways in which a tuner could go about curing a suspension problem, each of which would have slightly different effects. For example, the balance between oversteer and understeer, one of the most commonly tweaked issues a car has, is determined by the interaction between the tires, springs, and antiroll bars, as well as the car's weight distribution.

Changing the amount of oversteer or understeer requires changing the slip angle—the difference between the angle where the tire is pointed and where the car is actually going. Slip angle typically increases with the weight that a tire supports and how hard the car is cornering, until the tire reaches a point where it can no longer grip the pavement at all. If the tire is no longer perpendicular to the road, this will also increase the slip angle and reduce the available traction. There are then two different ways one can change the amount of grip a car has and the "balance" of a car's handling—by changing the grip of the tires, or by altering the way a car shifts its weight while cornering.

Weight Transfer

When a car goes around a corner, its weight will shift to the outside of the circle. The amount of weight that shifts is determined by the location of the car's center of gravity, the weight of the car, and how hard you are cornering. The suspension cannot do much to change the amount of weight transferred from side to side, but it can determine whether the shifted weight ends up on the front or back wheels.

CENTER OF GRAVITY

The center of gravity is the car's center of balance. It is a single point in space that may or may not be located on a physical part of the car. If you put a jack under a car's center of gravity, and the floor pan there could take the force, you could lift the entire car off the ground while keeping it perfectly balanced on the point of the jack. If you could stand the car up on its tail, you could balance it on the spot of its trunk lid directly below the center of gravity. And if the car was lying on its side, you could find the spot on the door under its center of gravity and . . . well, you get the picture.

The mathematical way to describe the center of gravity is as follows. Start at the center of gravity and move forward. At every point in front of the center of gravity, take the weight of what is located there and multiply it by how far in front of the center of gravity it is. Add up all these numbers. Now repeat the process for the points behind the center of gravity and add up all these values. The totals for both directions will be equal. The totals to the left and right will equal each other, although they may not equal the totals forward and backward. The same holds true if you go up and down from the center of gravity.

While that is the mathematical definition, it isn't very practical for use in the real world. Finding the car's center of gravity from front to back or left to right is not too difficult, as this can be done by measuring the weight on the tires. Many racing shops have suitable scales for this, which is known as corner weighing. You'll want to make these measurements with the amount of gas in the car you typically race with and the driver in the driver's seat (or an equivalent ballast weight).

It's more common to express the location of the center of gravity in terms of percent. The weight on the front wheels comes first. For example, if you find that 60 percent of the total weight is on the front wheels (not uncommon in a front-wheel-drive car), you have 60/40 weight distribution.

For the distance from the left wheels, you can use the same formula, only substituting the track width (the distance between the centers of the wheels on each side) for wheelbase (the distance between the centers of the front and back wheels), and the weight on the left wheels for the weight on the front wheels. Of course, since you want the center of gravity to be in the middle of the car from left to right, you can just make sure the total weight on both right wheels equals the total weight on both left wheels.

If you are building a car to handle and want to relocate parts or add ballast, it is usually not a good idea to place any additional weight behind the rear axle. This weight acts like a pendulum when turning. When a car exits a turn, the weight behind the rear axle tends to keep going the way it was turning, requiring extra grip from the rear tires to prevent the car from swapping ends. This can make for somewhat troublesome handling. A classic example is the Porsche 911, where mounting the engine entirely behind the rear axle makes this car unforgiving to drive at its limits.

Roll stiffness is a measure of how the car responds when its weight shifts from side to side. If you had a car with no roll stiffness, pushing the body sideways would cause it to flop over until its suspension bottomed out on the side where you put the weight. It also would not shift any of the car's own weight from one side to the other. If you happen to need a number to describe roll stiffness, it is measured in inch-pounds per degree. If a car had a roll stiffness of 5,000 in-lb per degree and you could somehow connect a very large torque wrench to its front bumper, applying 5,000 in-lb to the torque wrench would make the car lean over by one degree. The roll stiffness resists the torque by applying more force to the side where the car is leaning. If the wheels on that car were 50 inches apart, applying those 5,000 in-lb would cause the car to move around 100 lb of weight by reducing the force on the side that moves up by 50 lb and increasing the weight on the side that moves down by 50 lb. Usually it is difficult to just sit down and calculate how much roll stiffness you need unless you are a suspension engineer or a glutton for punishment, so most of the time you will simply be aware if you need more roll stiffness or less of it.

Each end of the car has its own roll stiffness. This is the key to tuning a car to balance out its handling, because the difference in roll stiffness can move the weight from one tire to another. If you add more roll stiffness at the front without making any other changes, the car will put more weight on the outside front tire when you take a corner and less weight on the outside rear tire. This will make the front tires drift more and the back tires drift less, increasing the amount of understeer. The opposite is also true: Adding roll stiffness at the rear and not the front will increase the amount of oversteer. If you add an equal percentage of roll stiffness at each end, you will not change the weight transfer and will simply wind up with flatter cornering, which may be just what you want if you were happy with the balance of the car before.

Body Roll and Traction

This assumes that the body roll did not cause any problems by pulling the wheels out of alignment, which can have its own effects on a car's balance. In many cases, body roll will cause the wheels to no longer be perpendicular to the pavement. In this case, reducing the body roll will improve traction on its own. This is particularly noticeable on cars that have very different suspension designs front and rear, such as most muscle cars with solid axle rear suspensions. The solid rear axle keeps both rear tires straight up and down, while the front tires will partly roll with the body. In this case, the body roll itself will cause the front tires to have less grip than the rear ones, and adding roll stiffness will reduce this effect. Reducing body roll will both minimize these effects and help the tires get maximum grip.

Tire Basics

Tires are an absolutely critical part of suspension tuning, providing the grip necessary to hold a car on a curve, transmitting bumps from the road to the suspension, and creating the reason behind oversteer and understeer. The width, sidewall height, tread pattern, and type of rubber used for the tread all affect the way a tire will behave.

One way to illustrate a tire's traction is a graph called a friction circle. The tire has a limited amount of traction that can be used for accelerating, braking, cornering, or some combination of these. The friction circle is drawn by showing the amount of grip that a tire has in all directions. For example, a very sticky tire can apply 1,000 lb of force to accelerate with no cornering or 1,000 lb of force while cornering without any acceleration. This will have a friction circle that looks like a circle with a radius of 1,000 lb. Suppose we wanted to find out how much cornering force was available while the tire was applying only 800 lb while accelerating. You would find the spot on the circle corresponding to 800 lb of force to accelerate and find that the tire has 600 lb of cornering force to spare. The size of the friction circle will change depending on how much weight is placed on the tire, the condition of the road, and even the temperature of the tire.

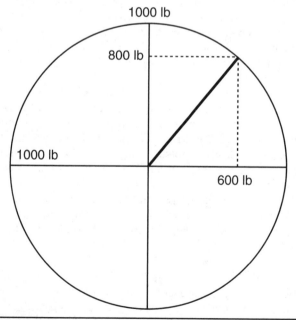

Friction circle for this imaginary tire.

Some tires may not have a friction circle that looks like a circle, either—a tire may be able to offer more grip while braking than while cornering, for instance.

It is possible to tune the tires you have by adjusting your tire pressure. A tire has a pressure at which it will give the most grip. Either increasing or decreasing the pressure from that point will result in less grip and more slip angle. So if a car understeers, it is often possible to change this by letting enough air out of the rear tires to bring their slip angle up to the same as the front tires. The downside of tuning with tire pressure is that the total grip available will be less. If you could tune out the understeer while keeping your pressure at the level to ensure maximum grip, you would have more grip available and be able to take corners faster.

If you are at an autocross, don't be afraid to experiment with extremely high tire pressures. Many tires can take as much as 60 psi for short lengths of time. Just be sure to return your tire pressure to a more reasonable setting for the drive home. Too low a pressure, however, is definitely not safe. Tires running at 20 psi will often feel like they are about to let go of the pavement, while dropping the pressure even lower can cause the tires to pop off the rim under hard cornering.

There are several areas to consider when choosing what size and sort of tires to use. Width is an important consideration. The biggest advantage to a wider tire is not that there is more rubber on the ground, as the area of the rubber actually touching the road is determined by tire pressure and the car's weight. Instead, the advantage is that the rubber that is in contact with the pavement is in a short band running across a wide tire

A tire for all seasons? The Kumho Solus KH16 is designed for dealing with rain and mud, and provides good durability. This means it won't be as sticky as an all-out racing tire. (*Photo courtesy Kumho*.)

At the other end of the spectrum, the Kumho Ecsta V710 is designed for maximum grip on a dry race track. (*Photo courtesy Kumho.*)

rather than a long band running across a narrow tire. This means that the rubber is in contact with the road for a shorter period of time and has less chance to overheat under hard acceleration, braking, or cornering. The disadvantage of wider tires is that they can be more likely to hydroplane if you drive through a puddle. Also, you will need to make sure that these wider tires will actually fit on your car before buying them. Tires that rub up against your suspension or hit the fender lip every time your car hits a bump will not last for very long.

The sidewall of the tire is the part between the tread and where the tire attaches to the rim. Sidewall height has a trade-off: A shorter sidewall will produce smaller slip angles for more control, while a taller sidewall will be better able to absorb bumps, not to mention protect the rim from getting bent or broken by potholes. Taller sidewalls also help increase traction in a drag race, as the tall rubber can "cushion" the tire tread from being violently broken loose by aggressive use of the clutch. Cars built for cornering tend to run larger diameter wheels and smaller sidewalls, while many drag racers prefer taller sidewalls.

The type of rubber and tread pattern are important in selecting a high-performance tire. Tires designed to last a long time are made of a hard rubber that holds up well but is not especially grippy. Race tires are often made of rubber that feels almost as soft as a pencil eraser and is so sticky that sand can stick to the tread. The tread pattern provides tires with channels for water to escape and groves that can grip into mud or snow, but

The Kumho Ecsta XS is somewhere in between, but balanced more to the performance end. (*Photo courtesy Kumho.*)

can reduce the total amount of grip a tire can apply on dry pavement. Some racing tires are known as "slicks" because they do not have any tread at all.

Tire Sizes

The modern sizing system for tires is a peculiar mishmash of millimeters, inches, and percentages, with several letters thrown in for good measure. They are relatively easy to understand once you know the code, however. Consider a street tire with the size designation P225/50ZR16.

The **P** in the code stands for passenger car. Many tires leave off this letter altogether. Occasionally, you will see a tire designation starting with some other letters, such as LT for light truck.

The **225** is the approximate width of the tire in millimeters. This is as measured at the widest point, so the tread width will be slightly less. Although the number is given to the nearest millimeter, in reality, the measurement may not be quite that accurate.

The **50** after the slash is what is known as the *aspect ratio*. This is the ratio of tire width to sidewall height, expressed as a percentage. On this tire, the sidewall height is 50 percent of the tire's width, so the sidewall of the tire is 112.5 mm tall. Tires with small aspect ratios, usually 50 percent or less, are known as *low-profile tires*.

The **Z** is the tire's speed rating. The letter stands for the maximum speed at which the tire can safely be driven (assuming you have a safe stretch of pavement with no speed limits). High-performance tires often have a V, W, Y, or Z speed rating. Some tires leave this letter off.

The **R** stands for radial. Some older tires use what is known as bias-ply or bias-belted construction, but virtually all new passenger car tires are radials. Drag racing slicks are an exception.

The **16** is the wheel diameter, in inches. To get the overall diameter of the tire requires a little math. First, convert all the measurements to the unit you want to use. Suppose we want to find the diameter of this tire in inches. We convert the width of 225mm to inches by dividing by 25.4, so the tire is 8.86 inches wide. The sidewall height is 50 percent of that, so it is 4.43 inches high. The overall tire height is the diameter of the wheel plus two sidewall heights, so this tire should be 24.86 inches tall. If you want the size in millimeters, use the width as it is, and convert the wheel diameter to millimeters by multiplying it by 25.4. This would show that the tire is approximately 631mm tall.

A common approach to choosing performance tires is what is known as plus sizing. This means selecting a wider tire that fits on a larger diameter wheel, but keeps the same total diameter as the original tires. A "plus one" tire would use a wheel that is an inch larger than stock, and also a little wider. Plus sizing lets the speedometer, computer, and other systems work as designed while letting you run less sidewall height. Installing tires with a different diameter than stock will make your speedometer read wrong at best, and make the antilock brake systems malfunction at worst.

Drag racing slicks use their own measuring system, which is based entirely on inches and requires considerably less math. A sample size would be 31 × 18.50 – 15LT. This tire would be 31 inches in diameter, about 18.5 inches at its thickest point, and fit on a 15-inch-diameter wheel. Note that these tires often have to use inner tubes, and many drag tires aren't street legal.

On rare occasions, you may see tires using other designation systems, such as a tire where the entire size description is "N50-15." The 15 is a 15-inch wheel, and the 50 is the tire's aspect ratio as with more modern tires. The *N* is somewhat more confusing. To make sense of this size, one would have to consult a chart. Tires that use this sizing system are either very old designs or modern copies built for exact restorations and period-correct hot rods. For those who want modern performance, such tires are best left in the museum.

Tire Types

Tires come in several varieties, each suited to a different purpose. Which tires to use will depend on how much performance you need, how long you expect the tires to last, the ride quality you want, and how much you can afford.

All-season touring tires are your basic, everyday tires. They are designed for a long tread life; a comfortable ride; and the ability to handle rain, mud, snow, and other hazards. Many accomplish this at a reasonable price, too. Their downside is that they are not designed for maximum traction. If you want your car to be an all-out handling machine, these just aren't adequate.

There are several other categories of "normal" street tires that offer more performance at a cost of ride comfort and tire wear. Some of these also trade away snow traction, and are described as "summer" tires. When eying the more expensive, maximum-performance street tires, keep in mind that the difference between them and a lower-priced choice may not be something that you can safely use on the street, and not enough to outrace someone on true racing tires. They may be the right choice if you want the ultimate true street car or compete in classes like SCCA Street Touring that ban the soft rubber used in all-out racing tires.

DOT-legal racing tires, often known as R-compound tires, offer about as much grip as possible for a street-legal tire. These tires will not behave quite like ordinary, everyday tires. They may not squeal at all until they let go, so you have to judge cornering by feel and not sound. They are sensitive to heat build-up and can harden up with time. These tires may not last more than 7,500 miles on the street, can be trouble in the rain, and should not be used in snow except in an absolute emergency. The advantage, however, is that they offer a level of grip you will not find in a normal tire. With the right suspension setup, a car on R-compound tires can corner at over 1 g.

All-out racing tires offer even more grip, but are not street legal. Many of them have no tread whatsoever, and can only be used on the driest of pavement. These tires tend to be quite specialized, as most are meant for either drag racing or road racing but not both. Drag slicks have sidewalls that "wrinkle" and wind up when the car is launched, preventing the power from being applied as suddenly and reducing the chance that the tires will come loose. A street enthusiast may keep a set of drag racing tires to put on at the dragstrip, but road racing tires are best left to hardcore racers.

Drag radials resemble R-compound tires, but are designed specifically for drag racing. Running lower pressure allows the sidewalls to wrinkle like drag slicks to get a considerable amount of grip when a car gets moving. Running normal tire pressures will make them work almost like R-compound racing tires.

There are some other specialized tire sorts for street cars. The temporary spare tire that comes with many cars today is meant to be light and cheap, and is not meant for anything other than getting 50 miles or so to the nearest gas station or tire shop to fix or replace the flat tire. They should not be used as a weight-saving measure on a drag car. Sometimes, however, the thin wheel used for a mini-spare can take a small conventional tire, giving you a cheap set of lightweight drag rims. Snow tires are meant for maximum traction on ice and snow, although they also grip ice-cold dry pavement better than summer tires.

Some newer tires offer run-flat technology that allows the tire to still be used after it loses its air pressure. These tires can be good for street use, but have extra stiffness that reduces ride comfort and have some quirks that make them less than popular with performance drivers.

Many owners of street cars that are occasionally raced will keep two sets of wheels and tires. One is for street use, and may be anything from cheap touring tires to high-performance street rubber. The other set is for race use, with R-compound tires, drag radials, or even pure race tires mounted on a set of lightweight racing wheels. This both avoids putting racing wear on (possibly expensive) high-performance street tires and keeps soft racing tires from racking up street miles.

Wheels

Wheels come in a wide variety of sizes and styles, but only use a few basic construction types. Picking the right wheel means balancing fashion, budget, comfort, and performance.

Increasing the wheel size can increase performance, but too big a diameter will only make the wheels heavier and cause both performance and ride quality to suffer.

The less expensive original equipment wheels are made from stamped steel. Steel wheels are quite strong and cheap, but heavier than your other options. Their strength makes them popular with circle track racers. Not only are steel wheels less likely to get damaged if you take the "rubbing is racing" mentality, but there are wheel shops that can fix these if you bend them, and for considerably less than it would cost to fix an aluminum wheel. Companies like Diamond Racing and Stockton Wheel Service offer steel wheels in an amazing variety of sizes to fit almost any car, making them a good option for a racer with an oddball car and a low budget.

Cast aluminum wheels are lighter, but more expensive. These wheels are made by pouring molten metal into a mold made from either sand or steel. Their low price and light weight make them a good choice for an enthusiast on a budget. Most factory alloy wheels are cast aluminum. On occasion, you will find wheels with cast aluminum centers and steel rims. The Cragar S/S is the most well-known example.

Billet wheels are made by taking a block of aluminum and machining it into the shape of a wheel center. This center is typically welded to a spun rim. These wheels are somewhere between cast and forged wheels when it comes to strength. The wide variety of shapes that are possible with billet makes these wheels very popular choices for show cars.

Forged wheels are made by taking a block of aluminum and squeezing it with a powerful press or hammer into the shape of a wheel. The center and rim are typically made separately and may be either bolted or welded together. Bolted together forged wheels are often called modular wheels, and come in a wide variety of sizes. Forged wheels are often the strongest and lightest wheels available, and not surprisingly, the most expensive. The strength does have a significant drawback: Squeezing the wheel into shape means that most of the stretchiness of the metal is gone. If a cast wheel is damaged, it will bend, and sometimes can be bent back into shape. Damaging a forged wheel will often cause it to crack or even shatter.

Some very high-end wheels are made from magnesium. The expression "mag wheels" originally described magnesium wheels, but later became a marketing term for aluminum wheels that imitated their look. Magnesium wheels are even lighter than aluminum, but more expensive.

Lightweight wheels sometimes are sold with warning labels. Some drag wheels are not for road racing or autocross use because they are not able to withstand sideways forces, only massive amounts of torque. Other wheels may be labeled "For racing use only," and are typically extra-lightweight wheels that can easily be damaged by potholes or other hazards encountered on the street.

Wire wheels are in a category by themselves, as many require attention and maintenance that other wheels do not need. While ordinary wheels are often all one piece, or at most three pieces of metal and some bolts to hold them together, wire wheels contain a hub with several parts, a separate rim, and dozens of wire spokes that can be individually adjusted. These wheels require periodic maintenance to keep the spokes tight. They also require special care when installing tires to avoid bending the spokes. If you want wire wheels, it is best to have a shop with experience in this sort of wheel maintain them for you. Wire wheels are usually not a good choice for performance use. Not only are they less strong than other designs, but many wire wheels require using inner tubes, which work best with tires designed for this use. Very few tires designed for good handling are meant to use inner tubes.

One feature that often shows up on wire wheels, but sometimes is used on other designs, is the knock-off hub. Instead of using a set of lug nuts to hold the wheel to the car, a knock-off wheel uses a single nut in the center to hold the hub of the wheel to a hub mounted on the car. Some of these nuts have small "wings" on them and use a lead hammer to tighten or loosen the nut. Others are octagonal and require a wrench. To transmit torque to the wheel, a knock-off hub has a set of splines between the two hubs. Tightening the nut is absolutely critical in keeping the splines from wearing out. Worn splines can often mean having to replace both parts of the hub, unless you want some very expensive machine work.

Wheel Sizing

The first thing people usually notice about a wheel is its diameter. Picking the right diameter is a mixture of performance and fashion. A larger diameter wheel can allow you to run lower profile tires for more precise handling, but too large a wheel will be so heavy it will slow you down, make the car ride roughly, and even create handling problems. A 20-inch wheel may weigh more than twice as much as a 15-inch wheel, and its tire is also likely to be heavier. For most performance-oriented cars, a wheel between 15 and 18 inches is often a good choice, with the 15-inch wheels offering lighter weight and a better choice of drag racing tires, and the 18-inch wheels offering a good choice of tires for precision cornering. Some newer cars will need larger than 18 inches to clear massive brakes.

Looks sometimes have their own rules. If you are not in search of ultimate performance, or if you have separate sets of wheels and tires for when you want to look your best and when you want to go your fastest, you may want a larger diameter than would be practical on a handling car. What looks right will depend on the size of your car. Twenty-inch (or even larger) wheels can look impressive on a Caprice, and might even look small on an Escalade, but would make a Mini Cooper look like a cartoon.

Width is the second most obvious dimension of a wheel. The width to buy is largely determined by the tires you plan to use, as wheels are usually the same width as the tire or only slightly narrower. In some cases you will need a narrower wheel to clear suspension components that the tires are not likely to touch themselves. Check with the manufacturer of your tires to see how wide a rim you will need.

The other dimensions are ones you will need to be sure your wheels fit your car correctly and work with your suspension. The first measurement you will need is the bolt pattern—the number of lug nuts you have and the diameter of the circle that they make. For example, the popular 5 × 4.5" bolt pattern has five lug nuts equally spaced on a 4.5-inch-diameter circle. If you have four or six lug nuts, you can measure the diameter by measuring the distance between the centers of two holes on opposite sides. If you have five lug nuts, measure the distance between the centers of two adjacent holes and multiply it by 1.7013. You will also need to make sure the wheel is designed for the sort of lug nuts you have, as lug nuts come in two sorts: cone and ball.

Most wheels are hub-centric. This means that there is a metal circle in between the lug studs that supports the wheel and makes sure it is centered correctly on your car. The hole in the center of the wheel should match the diameter of the circle on your hub. If the hole is too small, you will need to have a machine shop enlarge it. If the hole is too large, you can run a spacer ring (preferably metal) to center the wheels, or you can simply ignore this and bolt the wheel on anyway. This will make the wheel lug-centric, that is, positioned by the lug nuts. A lug-centric wheel can still be driven on the street, but may have more problems with vibration.

Backspacing and offset are two related dimensions that are important in making sure the wheel will fit. Backspacing is a measure of the distance from the inside edge of the rim to the wheel's mounting face—the point where the wheel mounts to the hub. Offset is somewhat more complicated. Imagine a line drawn right down the center of the wheel. If the mounting face is right on the line, the wheel has zero offset. If the mounting face is not on the line, the offset is the distance from this line to the mounting face. A wheel with the mounting face to the outside of the center line has positive offset, and a wheel with the mounting face to the inside of the center line has negative offset. (A few sources have been known to reverse these, but this is the more common convention.) Positive offset wheels are the most common, but a few older rear-wheel-drive cars and some cars set up to intentionally have the wheels stick out of the fenders will use negative offset. As a general rule, you should run an offset as close as possible to the offset of the original wheels and check to be sure how much backspacing your car can have.

Running wheels with an offset significantly different from the one your car was designed for can harm both handling and parts. Moving the center of the wheel will change the way the car steers. Also, the wheel bearings were designed for one particular offset. Moving the offset can increase the stress on these bearings. On some lowriders with wheels sticking out far beyond the fenders, the wheel bearings only last a few months.

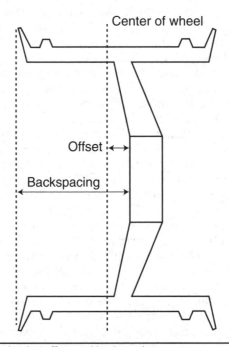

Cross-section of a wheel showing offset and backspacing.

Springs

Springs have several jobs in your car. The springs support the weight of the car and absorb the impact of bumps. Springs can affect many areas of a car's handling, from roll stiffness and the balance between understeer and oversteer to the harshness of the ride.

There are several varieties of spring. The most common sort is the coil spring, made from a large coiled steel rod. The metal in coil springs is often called *wire*, although this wire is much thicker than the sort of wire used for electrical work. Other sorts of springs include leaf springs and torsion bars. Leaf springs are made from strips of metal (or, sometimes, fiberglass) and usually attach to the rear axle on pickup trucks or older rear-wheel-drive cars. Torsion bars are sort of like an uncoiled coil spring, with one end attached to a suspension pivot point and the other end firmly attached to the chassis. The torsion bar absorbs bumps by twisting as the suspension moves. Torsion bars are most common on the front suspension of trucks, but appear on some performance cars as well, including early Honda CRXs, many Porsches, and older rear-wheel-drive Chrysler products.

Some springs are described as *progressive rate*. These springs will become stiffer the further they are compressed. This is done by causing some of the coils to bottom out before the spring is compressed as far as it will go. Progressive-rate coil springs usually either have different spacing between the different coils, or use tapered spring wire. The other sort of springs, known as *linear rate*, do not change their spring rate. Torsion bars are more or less linear, while some leaf springs are progressive. Progressive springs can give more comfort on the street, while linear springs can provide a race car with more predictable handling.

Some aftermarket springs come with "dead coils" that you can cut without affecting the spring's behavior. These coils are wound tightly enough to touch each other, so the dead coils do not move when the spring is compressed.

Swapping springs is a popular method of lowering a car. In many cases, it can be the only easy way. Lowering springs are nearly always stiffer, both to make the car corner in a flatter way and to prevent the suspension from damaging itself when it bottoms out. Some springs will also require cutting the bumpstops, which are rubber pieces that limit how far the suspension can travel when the car hits a bump. If you plan to cut the bump stops, be sure to leave enough material that you will not have other parts hit each other when the suspension touches the bumpstops.

Some people on a budget may try to lower a car by cutting the springs. This can result in springs that are too stiff for the rest of the suspension, or springs that are too soft to prevent the car from bottoming out. In especially unlucky cases, cutting springs can cause both problems at the same time. Still, this method can work if you don't get greedy. A typical rule of thumb is that you can get away with cutting one and a half coils off an original equipment spring if you are just trying to get a lower stance. While this may improve handling on some cars, many cars will handle about the same or slightly worse.

However, there are even worse ways to lower a car for those truly determined to ruin their suspensions. One popular bad idea was to heat the spring to make it sag, either by using a torch or putting it in the oven. Unfortunately, this weakens the metal of the spring. If you are lucky, the car will simply sag lower and lower as time goes on. If you are not, the springs may break while you are driving. Another bad idea is to put a set of clamps on the springs and simply drive with the clamps in place. While there are a few cases where properly designed clamps may work, poorly designed clamps can ruin the car's handling and may even cause springs to break.

Stiffer springs, in addition to preventing a suspension from bottoming out when you lower it, will add roll stiffness to the car. This is one reason why cars are likely to give flatter cornering when equipped with a well-designed set of lowering springs.

Installing springs is usually a relatively straightforward process, although it depends on how badly rusted your suspension is. Leaf springs can usually be removed and installed using just a set of wrenches. To install a coil spring will also require a coil spring compressor on most cars, unless you have bought a kit with a coil spring already wrapped around a brand-new shock absorber or strut. Torsion bars may require special tools—or some very creative improvising—to remove.

Shock Absorbers

Imagine gluing one end of a Slinky to the ceiling and the other end to a tennis ball. Raise the tennis ball up to the ceiling and release it. If there is enough room for the ball to drop before hitting the floor, the tennis ball will bounce up and down for a long time before coming to rest.

If a car had no shock absorbers, it would behave in the same way as that tennis ball. Shock absorbers damp out the vibrations caused by using springs to support a car. This not only makes for a much more comfortable ride, but allows the tires to be firmly planted on the ground and keeps the car under control. Cars with heavier weight or

An aftermarket strut for a Ford Mustang. Note the bolt brackets at the bottom and the pivot at the top. (*Photo courtesy Ridetech*.)

stiffer springs need more damping force. For performance driving, it is even more important to have precise damping than it is to have stiff damping. Shock absorbers are often called *shocks* for short, or dampers, which is technically more correct but usually only used by engineers and people who are British.

A strut is a special kind of shock absorber where the top end can pivot freely and the bottom end is bolted solidly to the suspension. Normal shocks pivot freely at each end. Some shock absorbers, and nearly all struts, run through the center of a coil spring, and may have the spring attached to them. A few cars, such as '80s-era Mustangs, use struts with separate springs.

There are several ways to build a shock. Virtually all designs on the market today use a piston to force oil through a series of valves. Gas shocks use nitrogen gas to keep a high pressure on the oil, preventing bubbles from forming in the oil and providing extra damping. Note that gas shocks often come with a strap around them or a wire clip. If you release this, the shock will expand to its full length. Sometimes this can be a problem when installing it yourself, depending on just how much pressure is in the shock. Twin-tube shocks have an inner tube for the piston and an outer tube that acts as an oil reservoir, while monotube shocks have a single tube with a very large piston. Monotube shocks offer good control in vehicles with limited suspension travel and resist heat build-up very well. Most racing shocks use monotube construction. Twin-

Cutaway of a monotube strut. (*Photo courtesy Ridetech.*)

tube shocks can work acceptably well in cars with somewhat longer suspension travel and can combine decent control with comfortable ride characteristics, as well as being able to still work if the outer tube is dented.

High-end shocks or struts often have adjustable damping. Some offer only one adjustment, while others can be adjusted for both compression (when the suspension moves up) and rebound (when the suspension moves down). How much adjustability you need depends on your preferences and how much experience you have in suspension tuning, as too many adjustments can sometimes give a novice too many items to worry about. Some high-end shocks for experienced racers offer four (or even more) types of adjustments.

A car that has been lowered with changes to the springs will virtually always benefit from high-performance shocks. Not only are the springs now stiffer and in need of more damping force, but the originals were probably not designed to work at the new ride height. Koni, Bilstein, KYB, and Tokico are some of the better-known makers of affordable performance shock absorbers, but there are several other manufacturers to consider if they offer shocks for your car, such as Penske, Ridetech, Dynamic, and Tien. Sometimes it is possible to install adjustable shocks meant for off-road trucks on passenger cars. High-performance models can cost from $100 each to four figures for all-out racing shocks. Most of the time, you get what you pay for. Recently, some companies have begun to offer packages that include both shocks and springs, which can not only save money, but often mean that the shocks and springs have been designed to work together.

Coil-over Conversions

Sometimes the word "coil-over" is used to describe any combination of a shock absorber or strut with a coil spring wrapped around it. Recently, however, this word has come to refer to aftermarket parts that fit around a shock and have a threaded spring perch to allow for easy changes in the ride height. Some coil-over kits work with an existing shock absorber or strut, and include appropriate springs. These can frequently be found for approximately $500 a set, sometimes less. More expensive coil-over kits include the shocks, adjustable perches, and springs in one unit. Expect to pay $1,200 or more for one of these kits from a top quality manufacturer.

There are several cautions about installing coil-overs. The first is that if you are installing them on a car that originally had the shock absorbers and springs mounted separately, you must first be sure the shock mounting point can take the additional weight. The second is that the car must have the weight at each corner compared when installing coil-overs or changing the ride height. Carelessly adjusted coil-overs can change the car's weight distribution, pushing down more weight on one corner of the car and consequently putting less weight on the opposite corner. The last caution is to make sure to get a new alignment after adjusting your ride height.

Air Bag Suspensions

Those who want maximum lowering in a car while still wanting practical street use may wish to consider an air bag suspension. Air bag suspensions appeared on heavy-duty trucks before being adapted for show cars, and many truck versions of air bags can improve load carrying capability as well. With air bags, you can let the air out to lower the suspension as far as it will go to give your car the slammed look and then reinflate the bags to drive the car on the street. Air bag systems typically include a compressor and controls to allow the driver to adjust the pressure. In some cases, air bags are sold as a bolt-on kit, while other systems need professional installation. Any quality air bag setup will either have tube-shaped air bags that surround the shocks, or use shock absorbers mounted separately from the air bags.

A well-designed air bag system will allow for comfortable driving and an impressive stance. Most air bags are not designed for maximum handling, but a well-designed system can get better handling than stock. They're generally not the way to win a serious autocross, though. The look and comfort also come with a price, as air bag systems are not cheap. Expect to pay over $1,500 for the parts alone.

A cheaper cousin of the air bag is the air shock. This is a shock absorber that you can pressurize with a tire pump to raise the height of the vehicle. These mostly fell out of favor about the same time the high-tailed stance went out of fashion. Using large amounts of pressure can damage mounting points if the air shock is installed in a car that did not originally use coils surrounding the springs. Air shocks do have one practical use, though, on drag cars. Mounting an air shock above whichever wheel loses traction first (usually the right wheel on solid axle rear-wheel-drive cars) can allow the driver to add air pressure above this wheel and preload the wheel to improve traction.

An air bag kit with struts. (*Photo courtesy Ridetech.*)

They usually make little sense on cars built to handle, as air shocks seldom provide the sort of damping you would find in a quality racing shock.

Anti-roll Bars

Anti-roll bars are one of the most effective tools for tuning the performance of your suspension. They are also known as anti-sway bars, sway bars, or stabilizer bars. As their name implies, these bars are designed to reduce body roll while cornering. Think of them as roll stiffness in a box. Most anti-roll bars take the shape of a steel rod with two bent ends. The center section attaches to the chassis with brackets, while the ends connect to the suspension with rods called end links. When the wheels on both sides move in the same direction, the bar rotates about the brackets and does not interfere with the motion of the suspension. When one wheel moves down and the other moves up, the bar twists, pushing the wheels so as to make the car lean less. Anti-roll bars reduce body roll while producing less ride harshness than stiffer springs. Since they make it easy to add roll stiffness, many tuners use them for adjusting the balance between oversteer and understeer.

The stiffness of an anti-roll bar increases as its diameter increases. For the mathematically inclined, the stiffness is proportional to the fourth power of the diameter. Replacing a half-inch anti-roll bar with a one-inch bar of the same shape will give you a bar 16 times as stiff. However, if two manufacturers offer different anti-roll bars of the same diameter for the same car, the bars may not be the same stiffness due to differences in the shape of the bar or the way in which they are mounted.

Some sway bars are adjustable. Usually this is done by having several mounting holes for the end links. The further along the bar the end links are mounted, the less stiff

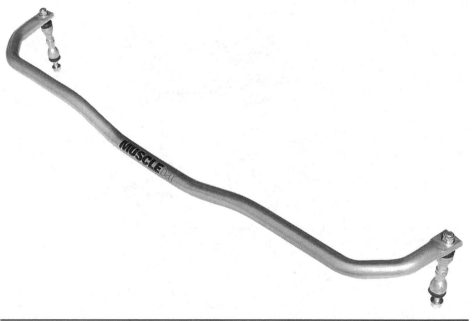

A bolt-on aftermarket anti-roll bar. (*Photo courtesy Ridetech.*)

the anti-roll bar will be. If you are using this sort of adjustment, always mount both end links in the same position. There are a few anti-roll bars that use different methods of adjustment.

The design of the end links is also important. Some end links have rubber bushings at each end. Using polyurethane bushings, bearings, or spherical rod ends on the end links will take some of the squishiness out of the end links and make the bar behave stiffer. In some cases, you may wish to add rubber to intentionally soften the anti-roll bar for tuning purposes.

One other feature to look for is whether the anti-roll bar is solid or hollow. A hollow roll bar will have the stiffness of a solid bar of the same diameter minus the stiffness of an imaginary bar that would occupy the hollow space inside it. Hollow bars are often much lighter for the same stiffness, but are somewhat bulkier. Since anti-roll bars often have complicated shapes, the way to calculate its stiffness is equally complicated. If you must know exactly how stiff one is, it is best to ask the manufacturer directly.

New anti-roll bars may cost anywhere from $120 to $300 each. In many cases, you will want to buy one for each end and install them as a matched pair. Installation may only require a set of wrenches. If your car was not originally equipped with anti-roll bars, you may need to drill a few mounting holes as well. Often, an anti-roll bar must be installed with the car supported by its wheels, either on ramps or on a lift, to prevent the bar from putting unequal loads on each wheel. Check with the directions that come with the bar.

The Chassis and Chassis Braces

There are several ways to build a chassis, the car's basic structure. The earliest cars used a framework to connect the suspension and drivetrain together, with a body bolted to the frame. This full-frame construction is very rare today, mostly used on trucks. Most cars today use unibody construction, where the body panels (usually the inner fenders, floorpan, roof, door frames, trunk floor, and rear fenders) carry the weight of the car directly. Some unibody cars also use subframes—small frameworks to support the engine and suspension—at each end. Unibody construction first became popular in the '60s, appearing on such classics as the first Mustangs. A well-designed unibody can often be stiffer than a full frame, but poorly designed ones may need reinforcement to get decent handling.

Many cars suffer from what is known as *chassis flex*. The body can bend slightly under hard cornering, making the handling feel less precise. Several sorts of braces can be installed to reduce chassis flex. Subframe connectors tie the front and back ends of the car together. Some subframe connectors can be bolted in, but welded-in subframe connectors are often significantly stiffer. A welded-in roll cage can have the same effect.

Strut tower bars and stress bars connect the right and left halves of the car at each end. These bars usually bolt in place. Some strut tower bars can be tightened. If you have such a bar, there is no need to tighten it any more than is needed to take any looseness out of its connections. The stiffness of a strut tower bar is determined almost entirely by its cross-sectional area and whether it is aluminum or steel. Steel is stiffer but heavier. When it comes to stiffness, the exact type of metal is not particularly important; a strut tower bar made from recycled beer cans will work as well as one made from top quality aircraft aluminum if they are the same shape and thickness. Bends in these bars weaken them, but adding welded-in metal triangles or other reinforcement at the bends can bring back some of that strength. One more thing to look for is to be sure the bar is not likely to hit any part of your engine, particularly if you have any unusual intake mods. Strut tower bars and stress bars are one category of mod where it is very easy for even a novice to judge the effectiveness.

There are other kinds of chassis braces available. Some cars benefit from having suspension mounting points or the steering box mount reinforced with welded-on gussets. A Monte Carlo brace is sometimes used to connect both tops of the shock absorbers to the firewall. Certain cars can receive help from bracing designed to correct specific problems.

Alignment

Wheel alignment is an important part of suspension tuning. Although it is possible to align your wheels at home with the right equipment, this is a task that is usually best left to a good alignment shop. A car should be aligned any time you change the ride height or replace any suspension part that cannot be replaced without disturbing the alignment. There are several settings that can be adjusted, but the most common are toe, camber, and caster. Some cars have more possible adjustments, such as ride height, while on other cars, it may not be possible to adjust camber or caster. Toe is always adjustable, at least on the front wheels.

Toe is the angle a wheel is pointed compared to the center line of the car if you look at it from the top. Toe can be measured in degrees, but it can also be determined by measuring the difference in the distance between the front edges of the tires and the distance between the back edges. Although the toe should be close to zero while the car is in motion, it is often necessary to set some amount of toe-in (or, on front-wheel-drive cars, toe-out) to compensate for the forces that act on the wheel while the car is in motion. Often, a small amount of extra toe-in will make the car feel more stable and react more slowly, while a little extra toe-out will make the car unstable. Way too much toe-in can also make a car unstable.

Camber refers to whether the tires sit perpendicular to the road. If the tops of the tires lean in toward the center of the car, the car has negative camber. A car with positive camber will have the tops of the tires leaning outward. Camber often changes as a wheel moves up and down. A graph of the camber as determined by the height of the wheel is known as a camber curve. Ideally, the outside tires will be dead perpendicular to the ground while cornering as hard as possible. Many cars will need to have the suspension set to a small amount of negative camber to achieve this, which can reduce traction while accelerating or braking in a straight line.

The camber setting recommended by the manufacturer may not be ideal for performance driving, particularly with a lowered car. One common test is to check how the tires are wearing out. If the tires wear more on the outside edge than the inside, the car has too much positive camber. If the tires have more wear on the inside edge, the camber is too negative. If the tires wear on both edges but not the center, the tires are underinflated. Wear in the center with little wear at the edge means overinflation.

Many racers use a special thermometer-like device called a *tire pyrometer* to conduct tests faster and with less wear on their tires. To use one, warm up the tires by driving the car quickly around a racing circuit, and measure the temperature at the inside and outside immediately after the laps. Too much negative camber will cause the inside to be hotter than the outside, and too much positive camber will cause the opposite. If the temperature is the same at the middle as well as both edges, the car not only has the correct camber setting, but this also indicates that the tire pressure is just right for maximum grip.

Caster is somewhat trickier to understand. When you steer a car, the wheels pivot about a fixed line. This line usually does not point straight up and down, but is at an angle. Caster is the angle between this line and vertical as viewed from the side of the car. If the line tilts backward, the car has positive caster. A few cars have negative caster, where the line is tilted in the opposite direction.

Positive caster will make the wheels self-centering; that is, the alignment will cause a force that steers the wheel back to the center when the car is in motion. Too much caster can make the steering feel heavy and the car transmit more bumps to the steering wheel. Too little caster will make the car unstable and make the steering feel numb.

Some cars will not have a built-in way to change the camber or caster. Often, this can be fixed with the right parts. One of the most popular devices is a slotted plate known as a camber plate that mounts to a suspension pivot point, usually the top of the strut. In other cases, the adjustment can be made by using special offset bolts or offset bushings. These are often called *crash bolts* or *problem solver bushings*, because they are usually used to align cars after a crash. In some cases, these aftermarket parts may be necessary even on cars where the suspension was adjustable from the factory in order to align a lowered car or reach an alignment setting beyond the original range of adjustment.

Suspension Designs

The design of a suspension needs to accomplish several goals. The suspension must allow for the wheels to move up and down while keeping them aligned with the road. It also must provide a way for the springs to support the car and the shock absorbers to control its motion. Last, the suspension should do this with as little unsprung weight as possible. Unsprung weight is the weight of any of the suspension parts that are not held up by the springs. Wheels, tires, and brakes are all examples of unsprung weight. When the car hits a bump, the momentum of the unsprung weight carries it upward and can only be stopped by the springs and shocks. The force needed to stop the weight is transmitted to the rest of the car through the springs, making you feel the bump harder on a car with more unsprung weight. This also means that for a brief moment, the wheels are not as firmly in contact with the ground as they could be. A car with high unsprung weight will not only ride more harshly on rough pavement, but it will not be able to handle as well.

When comparing different types of suspension, please note that the comments about the pros and cons are only general tendencies. A good designer can minimize the problems that can come with a basic suspension type, and good tuning can make a problematic suspension design still handle relatively well. Making the most of a suspension type is more important than starting with the "best" type of suspension. While a '57 Chevy uses unequal-length control arms and an AE86 Corolla GTS uses "inferior" McPherson struts in front, the Chevy is not going to have much of a chance beating the Toyota on an autocross course even if you equipped it with modern tires and shocks.

The oldest design is a solid axle. Solid axles have several good points. Since the axle does not lean along with the car while cornering, it keeps the wheels planted squarely on the road at all times. A solid axle suspension can be designed so that it will lift the back end of the car under acceleration, helping a rear-wheel-drive dragster shift its weight to the driving wheels. The drawback of a solid axle is that the heavy axle is completely unsprung weight. Solid axles are most familiar on traditional muscle and pony cars, but are also used on some less expensive front-wheel-drive cars such as Chevy Cavaliers, Chrysler K-cars, most Volkswagens, and recent Nissan Sentras. An axle that is also part of the drivetrain is called a live axle.

The solid axle needs to be connected to the rest of the car. On cars with coil springs, this is done with a series of bars with bushings at each end, known as links. A bushing is a thick tube made of rubber or plastic that is pressed into a hole on the link and allows a rod or bolt on the chassis to pass through it, making a somewhat flexible pivot. The most common arrangements use four links, and are appropriately known as four-link suspensions. Some cars may use only three, and others use a very long link called a torque arm combined with two short links. Front-wheel-drive cars with solid axles often use a cruder suspension that has only two links that are free to pivot at the end where they connect to the body but are welded or bolted solidly to the axle. If the car has leaf springs, the springs themselves usually take the place of the links.

Solid axle cars sometimes use what are called *lateral locating devices*. This is a fancy way of saying that the device prevents the axle from moving sideways. The simplest is a Panhard rod, which is simply a very long rod that runs horizontal and parallel to the axle, connecting the axle to the chassis. This does not do a perfect job of keeping the axle from moving sideways, since the end of the rod does not move perfectly up and down. A more effective lateral locating device is a Watts link, which uses an arrangement of three bars to restrict sideways movement. Many original equipment Panhard rods are not as stiff as they could be, and can be replaced with a stiffer version to make the suspension more precise. Solid axle suspensions with diagonal links generally do not need a lateral locating device.

Four-link suspension for a solid axle. (*Photo courtesy Ridetech.*)

A few front-wheel-drive cars use a torsion beam axle for their rear suspension. This resembles a solid axle, but the wheels are mounted on arms that extend out behind the axle. The axle twists in much the same way as an anti-roll bar.

Most other suspension designs are known as independent suspensions because the movement of one wheel does not cause the other wheel to move. Independent front suspensions are by far the most common sort of front suspension, with solid axle front suspensions being limited to street rods, antiques, trucks, and off-road vehicles. Independent rear suspensions are not as universal, but show up on most front-wheel-drive cars and the more expensive rear-wheel-drive ones. Independent suspensions have less unsprung weight for a smoother ride, and a well-designed suspension can keep camber under control almost as well as a solid axle.

Unequal-length control arms are one of the most popular independent suspension designs. The wheel and brakes are attached to a spindle, which is equipped with two ball joints, one above the other. The ball joints are pivots that allow for a wide range of motion in all directions. Each ball joint is attached to a control arm that runs inward and pivots at a bracket on the frame or body, usually using two bushings. Each spindle is also connected to a part called a tie rod to keep the wheel pointed in the right direction. On the front end, the tie rod runs to the steering mechanism. On the back end, the tie rod is usually attached to a point on the chassis, but some rear suspensions replace the ball joints with bushings and do not use tie rods. This design is sometimes also called a *double wishbone suspension* because of the shape of the control arms.

The unequal length of the control arms allows the designer to control how much the camber changes when the body rolls. Compared to other independent suspension

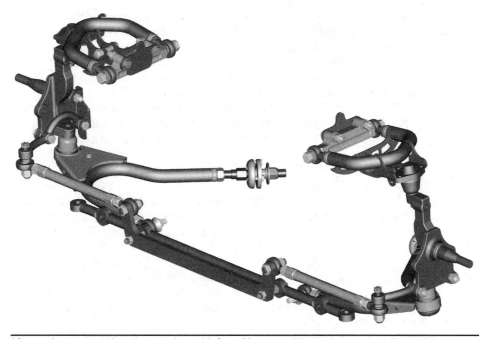

Aftermarket unequal-length control arm kit for a Mustang. (*Illustration courtesy Ridetech.*)

designs, this suspension type places the fewest restrictions on the suspension design. There is a reason this sort of setup is used on formula cars.

Strut-type suspensions are the other common type of independent suspension. These use a lower control arm and spindle-like, unequal-length control arms. Instead of attaching the top of the spindle to a control arm, it bolts to a strut that holds the spindle upright. Strut-type suspensions typically free up more room in the engine compartment and cost less than unequal-length control arms. The disadvantage is that struts often do not control camber as well on cars with significant amounts of body roll.

Many unequal-length control arm and strut-type suspensions do not use one piece, A-shaped lower control arms. For various reasons, most of which have to do with lower costs, the lower control arm may simply be a straight link with one bushing connecting it to the chassis. This requires another suspension part to be added to prevent the lower control arm from wiggling back and forth. This is usually known as a *strut rod* or *radius rod*.

Some rear-wheel-drive cars use a semi-trailing arm rear suspension. This attaches the wheel to a single large arm on each side that pivots diagonally. Some designs use a tie rod to control toe. Others have the wheel mounted solidly to the semi-trailing arm, making the movement of the suspension also steer the wheels. If the toe angle in a semi-trailing arm suspension is not kept under control, the suspension may need to be stiffened considerably to keep this from causing handling problems. Trailing arm suspensions are somewhat less common; these use an arm that runs in parallel to the car's direction of travel instead of pivoting sideways. This does not have problems with toe angle changes, but it keeps the wheel always tilted at the same angle as the car body, which can hurt traction if the car leans significantly in a corner. Confusingly, the word "trailing arm" is also sometimes used to describe a link in a rear suspension that works in a similar way to a strut rod.

On some cars, it is possible to buy aftermarket control arms or suspension links to correct for shortcomings in the original design. These parts may be designed to minimize the effects of body roll on camber, to change the alignment beyond settings possible with the original parts, or to prevent the suspension from binding. Binding is a problem that happens when the suspension moves in such a way that the bushings are compressed almost to their limits and start limiting the suspension travel as if the springs had suddenly become much stiffer. Aftermarket control arms and links may also be adjustable, allowing for changes in alignment. In some cases, it is also possible to install stiffer tie rods, which can make for more precise steering control.

Traction Bars

Traction bars are used on rear-wheel-drive cars with leaf springs and live axles in the rear suspension. Under hard acceleration, the front section of the leaf spring can twist into an S-shape, pulling the axle upward and reducing traction. This is known as axle windup. Axle windup can be partly reduced by making the front half of the leaf spring shorter and stiffer, but another way around this problem is to use traction bars. There are three common types of traction bars on the market.

The simplest type of traction bars are also known as *slapper bars*. A slapper bar is a solid piece of rectangular tubing with a rubber bumper at one end and a mounting bracket at the other. This bar mounts to the rear axle under the leaf spring. If the spring twists, the rubber bumper touches the chassis and limits how much the axle can wind up. Some cars use a variation of this by mounting a rubber bumper called a pinion snubber in the center of the axle to the front of the differential. For drag racing, you can add an adjustable pinion snubber and set it much closer to the floorpan. The disadvantage of the slapper design is that the axle already needs to wind up a little before the traction bars take effect.

Another sort of traction bar uses a solid link with pivots at each end. One end is connected under the point where the leaf spring connects to the axle, and the other end mounts to a bracket on the rear subframe. This design works well if the leaf spring is already quite flexible, and has the added advantage of preventing axle windup under braking too. This sort of traction bar should never be used on cars where the front half of the leaf spring is already quite stiff, such as most Chrysler products from the '60s through the '80s. Installing them on this sort of setup will make the suspension bind and can damage the springs or suspension mounting points.

The newest sort of traction bar is designed to attach to the pivot point of the front half of the axle. This type is stiffer than a slapper bar, but does not create problems for cars with stiff leaf springs.

One other trick that drag racers frequently use is to unclamp the spring leaves on the rear half of the spring while leaving the front half clamped firmly. This makes the spring less stiff in extension, making it easier for the suspension to lift the back end into the air under hard accelerating. This transfers more weight to the rear wheels, allowing for more traction.

Bushings

Most cars leave the factory with rubber bushings in their suspension. Rubber has several advantages. It's cheap, absorbs bumps well, lets the suspension turn freely, and won't bind easily. Unfortunately, it also makes the suspension somewhat squishy. This can make the suspension feel slightly vague and contribute to wheel hop, especially in front-wheel-drive cars.

Replacement bushings can be made from several different materials. Polyurethane, the most common material used, is a synthetic rubber that is as much as 100 times as stiff as natural rubber. In some cases, polyurethane will make a suspension noticeably harsher, but in some cases you may not even notice. Its downside is that its high level of friction does not allow parts to slide across it as easily. In some cases, a badly designed set of polyurethane bushings can "grab" the suspension and make it behave as if it were running a second set of springs—a much stiffer set, making the suspension virtually untuneable. If the bushings are not properly greased (some of them have built-in graphite lubrication for this) they are likely to squeak. Also, if a suspension design has problems with binding, polyurethane's additional stiffness will make matters worse.

Those seeking a stiffer suspension can use bushings made from Delrin. Delrin is an extremely stiff plastic with low friction. These bushings must be precision machined because they cannot twist at all. These can make a ride more punishing than a good set of polyurethane bushings. However, it will not be anywhere near as harsh as the results of a poor quality set of polyurethane. Delrin will give a car even more of the benefits of polyurethane while avoiding some of its drawbacks.

For the truly hardcore, you can replace the bushings with metal bearings. Bushings that only rotate can be replaced with conventional roller bearings, while bushings that move in several directions must be replaced with Heim joints or spherical bearings.

These offer very low friction and no "squishiness," but do not absorb impacts like rubber, and need to be greased from time to time. Some designs will also wear out in an unacceptably short time if used on the street, so check with the manufacturer before putting these on a street car. If your suspension has any problems with binding, you should not install Delrin bushings or metal bearings unless you can make changes to prevent these problems. Delrin would be a very bad choice for an otherwise stock rear suspension on an '80s-era Mustang, for example.

Replacing bushings often requires a press and may need to be done at a machine shop. One way you can minimize downtime while changing bushings if your car is relatively common is to buy a spare set of control arms, either new or from a junkyard. This avoids spending time with the car partially disassembled, waiting for machine shop work. Be very careful if using control arms from a car that has been sent to the junkyard because of a crash. They may be warped or have cracks, some of which may not be visible.

Drivetrains and Handling

Which end drives the car can influence the way the car handles, at least when the car is accelerating. Acceleration may use up part of the tire's available grip, which can cause the driving wheels to lose traction. Give a rear-wheel-drive car too much throttle coming out of a low-speed corner, and it is likely to oversteer or spin out. Doing the same in a front-wheel-drive car is likely to cause the tires to spin and prevent the car from responding to steering input until you let off the throttle. An all-wheel-drive car will often be able to put down more power in this situation and exit the corner with more speed, but giving it too much throttle may make either end come loose.

However, this is not the only effect that comes into play when changing the speed while cornering. Acceleration shifts the weight of the car rearward, which will plant the rear tires and give them more grip. Slowing down will shift the weight forward, which up to a point will cause the car to oversteer. This means that there are times when giving a front-wheel-drive car a small amount of extra throttle in a turn will decrease understeer. It also means that there are situations where accelerating in a rear-wheel-drive car may actually make it stick more. There are even perverse cases with rear-wheel-drive cars, often when they start to fishtail, where giving it more gas will cause the tail end to slide, but lifting your foot off the throttle will make the car spin! If you find yourself trapped in this situation, the best you can hope for is to keep an even throttle and straighten out the car, at which point you can safely hit the brakes.

It is important to remember that which end of the wheels drive the car is not important while cornering and braking, and has practically no effect while cornering at a steady speed. The drivetrain effects mostly come into play while accelerating out of a corner.

The choice of differential can also influence the way a car handles. An open differential will put an equal amount of thrust on each side, so it will have little influence on the way a car handles. Most performance differentials will transfer more power to the wheel with more traction, which is the outside wheel when cornering. This extra thrust on the outside will contribute to oversteer. If the differential is also partly locked while the car is not under power, it will produce understeer while coasting or braking by creating drag on the outside wheel. Detroit Locker differentials, however, can have the opposite effect: When they unlock, they drive only the inside wheel. A differential of this type will cause understeer when cornering under low power, but can create an abrupt change in handling characteristics when enough power is applied to cause it to lock. This makes the Detroit Locker better suited to drag cars that see street use than all-out handling cars. Some cars use a computer controlled differential to adjust the car's handling.

Locked or partially locked differentials can create even more nuisances in a front-wheel-drive car. When the wheels should be turning at different speeds and do not, the halfshafts can wind up like springs and then release this energy in ways that transmit a lot of vibration to the steering wheel. The wheels will also resist turning into a corner. For a front-wheel-drive car built to handle, it is best to use a differential that acts like an open differential when you have your foot off the gas.

Sometimes, the drivetrain not only changes how the car handles, but can actually change the direction the wheels are pointed. This effect is called torque steer, and can be very troublesome when it happens. Torque steer is most common on front-wheel-drive cars, but rear-wheel-drive cars with independent rear suspensions are not always immune either. Torque steer can be caused by the force of acceleration causing the bushings to bend and move the wheel out of alignment, or by the torque of an axle halfshaft acting directly on the spindle. If the halfshaft is completely perpendicular to the tire, all of the torque transmitted by the shaft will go into rotating the tire. If the halfshaft is pointed at an angle to the tire, however, a part of its torque will go into trying to rotate the spindle instead of the tire. If the halfshafts are the same length on each side, this effect will be roughly equal from side to side and effectively cancels itself out. If the halfshafts are of different lengths, or if the car is leaning as it goes around a corner, the effect will be much more noticeable. In some cases, you may be able to convert a car with unequal-length halfshafts to equal-length halfshafts using junkyard parts. Stiffer bushings and beefier tie rods can also help minimize torque steer.

Lowering Done Right

Many beginners want to start off work on their suspension by lowering their cars. Some want to lower their cars, hoping it will improve handling, while others simply want the slammed look. There are several ways of lowering a car, depending on your goals.

By itself, lowering can make an improvement, although it will not give you the same grip as R-compound tires or reduce body roll as much as adding stiffer springs or anti-roll bars will. Lowering will drop the car's center of gravity, meaning less weight will shift from side to side while cornering. This helps somewhat, but lowering also can create several handling problems. The control arms will be angled differently, which will put the car on a different part of the camber curve and may reduce the car's grip. The shocks may not work as well at the new ride height. The suspension may bottom out on bumps or corners, or the car may lose so much ground clearance that it gets stuck on speed bumps or hits road debris. Doing the lowering incorrectly may create even more problems.

Note that all of these issues are described with "can" or "may," not "certainly will." Some cars with well-designed suspensions will work quite well even if they have been dropped several inches. Other cars will start to run into trouble if they are lowered past an inch. If you do not know for sure which category your car falls into, it is best to play it safe when it comes to lowering. Usually, cars with unequal-length control arms or solid axles can take more lowering than ones with struts.

There are several options for how to lower a car. For cars with coil springs, the most common methods involve changing either the springs or the way the springs are mounted. Lowering springs are fairly popular mods, but in some cases it is possible to cut an inch or so off the spring to lower the car. Springs can be cut with a hacksaw, but this takes time. A cut-off wheel or cutting torch is faster. Some sorts of springs are not good candidates for cutting for anyone who is not very familiar with spring design. If the coil spacing changes from top to bottom, the diameter of the coils change, or the spring is made from tapered wire, it should be left alone unless it has "dead coils" put there specifically to let you trim it. Keep in mind that in most cases, spring cutting is simply an appearance mod and may not improve the handling. If you are lowering the car by making any changes to the springs, even if you are simply cutting a coil off, it is a very good idea to upgrade your shocks. Stiffer springs with original equipment shock absorbers will often make for an underdamped car.

On some cars, you can use a kit to move the spring's mounting points. Some aftermarket struts or shock absorbers may come with lower spring perches. Adjustable coil-over kits allow you to move the spring perches using a special wrench to set a wide variety of ride heights. On a few cars, you can even move the spring perches with bolt-on spacers.

Suspensions that do not use coil springs have other ways in which they can be lowered. Torsion bar suspensions often have an adjustment screw that lets you adjust the ride height as if the car had a set of adjustable coil-overs. In some cases, the ride height on a torsion bar suspension may be changed by removing the bar and turning it

A wrench for adjusting coil-overs. (*Photo courtesy Ridetech.*)

Drop spindles, like this Ridetech kit for '55–'57 Chevys, let you lower a car without making the ride stiffer. (*Photo courtesy Ridetech.*)

in its mount. Usually, you can find information on how to adjust the ride height in a good repair manual if your car has torsion bars. You can get lowering springs for leaf spring suspensions, but another way to lower them is to install inexpensive lowering blocks between the spring and the axle. You can also have a spring shop de-arch leaf springs or relocate the mounting points. Remember if you are lowering a car without adding stiffer springs, make sure not to lower it far enough for the suspension to bottom out or the car to scrape the ground.

There are other workable ways of lowering a car. On some cars, you can install drop spindles, replacement spindles that lower the ride height without making any changes to the rest of the suspension. These let you keep the same ride characteristics and simply have a lower stance. Air bag suspensions are popular for cars built to both cruise and display at car shows. They combine a comfortable ride with the ability to let all the air out and drop a car to the ground when it is parked.

Tuning Suspensions

Armed with a good understanding of how the suspension components interact, you can begin to find ways to modify your car for better handling. This begins with examining the problems your car has with its handling and finding ways to correct this. In some cases, you will need to ask yourself why the car left the factory the way it did before making a change. In the case of an older car, this may be because it was designed for tires with much less grip than modern radials. Family cars are often designed with low cost and a soft ride taking priority over ultimate handling. At the other end of the extreme are cars like the NSX, Viper ACR, and Porsche 911, built with a "money is no object and handling is a key goal" mentality. Lower-priced performance cars often fall in between these extremes and may have a few corners cut where they can be improved

Good suspension tuning requires extensive testing. (*Photo courtesy Edelbrock.*)

or compromises made for the sake of street driving. Handling is one area where the saying "If it ain't broke, don't fix it" applies.

Much of the tuning requires changes to the car's roll stiffness. Since both the springs and the anti-roll bars determine roll stiffness, you can swap in a stiffer version of either one to add stiffness or a softer part to decrease it.

Problem: Not Enough Grip

This one is straightforward: The car simply cannot take a corner as fast as you want or as fast as your opponents are going because its tires start to let go too soon. The first thing to do about this is to get stickier tires. The other things to do if your car rolls a lot is to increase the roll stiffness so the tires stay flatter on the ground or lower the car to keep the weight from shifting as much. Alignment can help, too, as optimizing your camber settings will keep the tires flatter on the pavement.

Problem: Too Much Body Roll

A car that leans too much may not only suffer from less grip, but will feel out of control and inspire less confidence. This can be cured with either stiffer springs or stiffer anti-roll bars. If the car's handling is already relatively well balanced otherwise, you will want to stiffen both ends at once. Installing stiffer anti-roll bars often makes for a softer ride than stiffer springs and is less likely to require new shocks.

Problem: Understeer

If the suspension design is similar at each end, understeer can be reduced by adding more roll stiffness to the rear suspension or reducing the roll stiffness in the front suspension. On rear-wheel-drive cars with a solid rear axle, the understeer may partially be caused by the camber change at the front wheels. Putting a stiffer anti-roll bar or springs on the *front* of the car can reduce understeer in this case, but only to a point. Beyond this point you will need to add roll stiffness at the rear instead. This is why many muscle cars run a stiffer anti-roll bar in the front than on the rear. Nose-heavy front-wheel-drive cars, by contrast, run stiffer anti-roll bars at the rear.

Tire changes can also reduce understeer. If you have installed larger tires on the back than on the front on a car that originally used equal-sized tires on all corners, this combination is a surefire recipe for understeer. This is also true if you have a combination of 14-inch stock steel wheels and 60-series out front with 17-inch wheels and low-profile tires on the back, even if the tire width is the same. Placing grippier tires on the front than on the rear will generally reduce understeer or create oversteer. So will running more tire pressure in the front than in the back, which can be a useful tuning technique if you need last-minute adjustments.

This problem can also be reduced by weight distribution and downforce in some cases. Lightening the front end will mean the front tires have to provide less cornering force, which can improve matters. Creating more downforce on the front of the car with an airdam or splitter can help at high speeds, but will not be of much help in an autocross.

If the car only understeers when entering or exiting a turn, the problem may lie with the shock absorbers. You can stiffen the rear shocks or soften the front ones to make the car turn in faster. If you have a set of double-adjustable shocks, it is best to adjust only the rebound to reduce oversteer. Often, increasing the rear rebound stiffness is the best adjustment to reduce understeer when entering a corner. Increasing toe-out on the front wheels may also help.

Problem: Oversteer

Oversteer is often cured by taking the opposite of mods that cure understeer. Increasing front roll stiffness, decreasing rear roll stiffness, or putting grippier tires on the rear can all be effective treatments. If a car oversteers more at high speeds, this may be caused by aerodynamic lift. Installing an effective rear wing can reduce high-speed oversteer caused by lift. This is the reason why road racers often run rear wings, even on front-wheel-drive cars.

As with understeer, a case of oversteer that only happens at the beginning or end of a corner is likely to be a shock problem. Usually, the best way to cure this is to reduce the rebound damping at the rear, although sometimes you may need to add more front rebound damping instead.

Problem: Suspension Feels Vague, Bouncy, or Floaty

These symptoms often point to problems with the shock absorbers, particularly if they happen on rough pavement. The symptoms may vary from a car bouncing violently from side to side on rough pavement, to the suspension feeling like it has not "settled" for two or three bounces after you hit a bump. The shocks may be worn out, too soft for the springs you have chosen, or simply too soft for spirited driving.

It is also possible to set the shocks to be too stiff, a problem that usually will not show up on an unmodified car. This not only creates a harsh ride, but can prevent the tires from staying firmly in contact with the pavement. A car with excessively stiff shocks may have the tires suddenly lose traction during hard cornering.

Problem: Car Feels Unstable on Rough Pavement

If the shocks are in tune, this points to a problem with bump steer. You can check for this by supporting the car on jackstands and using a jack to move the suspension through its range of motion, checking if the toe angle changes. If this is the case, you will usually need to adjust the suspension so the tie rod is at the same angle as the lower control arm at normal ride height. You may have to modify the steering knuckle or move the steering rack with spacers to achieve this.

Problem: Car Feels Unstable, Even on Smooth Pavement

Most sorts of instability can be traced to alignment problems or worn-out suspension parts. It can be difficult to tell which is the problem without inspecting the suspension, but once worn parts are found, the cure is relatively straightforward. Alignment problems that cause instability are usually caused by too much toe-in or sometimes toe-out. Too little caster can also contribute to instability. A good alignment shop will be very helpful in diagnosing this problem.

Some drag cars may become unstable when you hit the brakes at the end of the run. It can feel as if the car is trying to swap ends. If this happens only at high speed, the problem may be aerodynamic lift. In this case, you may need a rear wing to keep the tires planted. In some cases, an adjustable proportioning valve in the brakes may also solve this problem without adding the sort of drag that a wing would cause.

Problem: Car Does Not Keep All Four Wheels on the Ground While Cornering

Many lightweight, front-wheel-drive cars will lift the inside rear tire on an autocross course. Lifting the front tire happens with Porsches and some heavily modified muscle cars. If this happens, the only way to keep the tires planted is to add more roll stiffness at the end that can keep both wheels on the ground. Unfortunately, many cars that lift their rear wheels will also understeer, and stiffening the front end will make this problem even worse. The solution is to stiffen the rear anti-roll even more than the front to compensate for this. With especially nose-heavy cars, you may not be able to dial out all of the understeer.

Problem: Car Bottoms Out

A car may bottom out while going over bumps, or it may bottom out while cornering. If it bottoms out while cornering, the end that bottoms out will suddenly become much stiffer and the car will immediately lose traction at that end. The usual solutions are to raise the ride height or install stiffer springs. If the car only bottoms out on corners, it may be possible to cure this problem with stiffer anti-roll bars instead.

Problem: Not Enough Traction off the Line

Drag racing traction can be improved by addressing four different problems: differential behavior, wheel hop, tire grip, and weight transfer.

If you have an open differential, having one tire lose traction before the other one will cause the wheel with less traction to spin and the other wheel to stay put. Installing a performance differential can avoid this problem, but if you do not have one available for your car, or if you simply want to try less expensive options first, there are ways to reduce this problem. Remember that an open differential feeds an equal amount of torque to each wheel. If you can balance the amount of grip each tire has available, you can launch harder without doing a one-wheel burnout. Shift more of the car's weight to the wheel with less traction. This can be done with suspensions that have an adjustable ride height, but can also be done by inflating an air shock above the wheel with traction problems by moving around parts like the battery, or by adding ballast. There are front-wheel-drive cars that can cover the quarter mile in ten seconds using an open differential and careful weight tricks.

Wheel hop occurs when parts of the suspension, such as bushings or leaf springs, distort under acceleration, pulling the wheels upward. The wheels temporarily lose traction and the suspension unwinds, planting the tires again, which causes the wheels to hop until the car finally gets moving or the driver backs off on the throttle. This not only costs you traction, but can break U-joints or other drivetrain parts. A few poorly designed suspensions will pull the wheels up under hard acceleration even if the force does not distort any parts. On rear-wheel-drive cars with leaf springs and solid axle rear suspensions, installing traction bars can minimize this. A rear-wheel-drive car with a solid axle and a link-type suspension can sometimes benefit from aftermarket links.

If your car has an independent suspension at the drive wheels, often the only thing you can do is to install stiffer bushings and hope for the best. Front-wheel-drive cars also benefit from stiffer motor mounts. In some cases there are kits specifically designed to prevent wheel hop on a particular sort of independent suspension. Some drag racing enthusiasts have gone so far as to convert a car from independent rear suspension to a solid axle, but this usually requires custom work unless the car could be ordered with either setup and you can bolt the axle right in. About the only car where this is an easy swap is the SN95-series Mustang Cobra.

When a car accelerates, its weight shifts toward the rear of the car. The higher the center of gravity, the more weight is transferred. This is good for a rear-wheel-drive car, but is a problem if the front wheels are trying to get the car moving. Rear-wheel-drive cars can increase weight transfer by using soft front springs and running front shocks with a soft rebound combined with stiff compression. If you have a solid axle suspension, you can tweak it using aftermarket links, traction bars, or leaf springs specifically meant for drag racing to lift the rear end under acceleration. Some drag cars will actually lift up at both ends when the light goes green. Cars with independent rear suspension and rear-wheel drive should use stiff rear springs and shocks that are stiff in compression.

Front-wheel-drive cars have the opposite problem. Letting the front end rise would further limit traction, so they benefit from stiffer front springs and plenty of rebound stiffness in the shocks. A rear suspension that compresses easily can reduce weight transfer. Sometimes front-wheel drag racers actually take "wheelie bars" meant to control wheelstanding rear-wheel-drive cars and modify them so they press down on the ground and lift part of the weight off the rear wheels.

A good drag racing tire has a wide tread, sticky rubber, and a tall sidewall that can twist and wrinkle to absorb the force of suddenly putting power to the ground. Drag racing tires often benefit from being warmed up by a short burnout, but street tires often grip better when cold.

Driving Classes

If you are serious about getting around a road course or autocross as fast as possible, one of the most important changes is not one you make to the car. Experience and good driving skills are even more important than car setup. Driving schools exist for many different budgets, from road race schools that last for days and cost thousands to single-day autocross classes that may cost only a few hundred dollars. This may seem expensive, but the skills you learn may cut more seconds off your lap times than any performance parts you can find for the price.

CHAPTER 10

Brakes

Optimizing Your Stopping Power

It seems like there would be no need to make compromises and trade-offs with brake upgrades. A car can't stop too hard, right? The problem is that a set of brakes meant for all-out racing might not stop very well on the street, while the brakes that stop the fastest on the street might quickly overheat and go dead on a race track. When planning brake mods, it is important to be realistic about how you will use your car to

Is it possible to have too much braking power? Not if you're building a car for all-out road racing and giving the brakes a massive workout. (*Photo courtesy Baer.*)

make sure you do not hurt your performance. If you try to solve a problem that your car doesn't actually have, you can end up creating new problems instead.

Tires and Brakes

Just like a car cannot corner with more grip than what its tires make available, the tires are the biggest limit to your braking system. If you try to stop with more force than your tires can put to the pavement, either your brakes will lock up and the car will skid, or the antilock brake system (ABS) will kick in and try to keep you from stopping any harder. The only cure is to install a set of stickier tires.

So, if the tires are often the limit, why try to add brake mods to get more stopping power? There are three things that brake mods can accomplish. First, if your brakes overheat, they may no longer have enough grabbing ability to lock up the tires. Stock brakes are intended for in-town driving and the occasional interstate panic stop, but few stock brakes were meant for slowing down a car on repeated laps around a racetrack at triple-digit speeds. Aftermarket brake mods can allow your brakes to not heat up as fast, or to still work even when smoking hot. Second, you can tweak the brakes to balance braking force evenly between both ends of the car, allowing you to get the most out of the tires you have. And third, upgrading your brakes can make them easier to control.

Road racing is one of the most demanding applications for brakes. The cars may not be going as fast as a dragster, but having to slow down lap after lap results in more heat build-up. (*Photo courtesy Baer.*)

Anatomy of a Braking System

Unless you are restoring an antique from the days of Model Ts, your car probably has a hydraulic braking system. With hydraulic brakes, the brake pedal pushes on a hydraulic cylinder, known as the *master cylinder*. This forces brake fluid through the brake lines to the brakes themselves, which on normal street cars sit just behind each wheel. A few Jaguars and other European cars use inboard brakes, where the brakes are attached to the differential instead of the wheels to reduce unsprung weight.

Brakes come in two varieties: disc brakes and drum brakes. Almost all cars since the mid-'70s use disc brakes on the front wheels, and high-performance cars often have them at the back too. A disc brake has a spinning metal disc called a rotor that attaches to the wheel or axle. A mechanism called a caliper grabs the rotor. Since metal grabbing metal would mean an unacceptable amount of wear, the caliper uses two brake pads designed to take the friction. Disc brakes are popular for high-performance use because they offer excellent cooling and grab in a predictable fashion.

Drum brakes are usually found at the rear of less expensive cars, although some cars from the '60s and earlier also used them at the front. A drum brake uses a pair of curved pieces of metal called brake shoes that are coated with the same material that is used on brake pads. A hydraulic cylinder forces these pads outward so that they push against a metal brake drum. Although the main reason manufacturers use drum brakes is a cost-cutting measure, they can give a performance advantage on a drag car. Drum

Disc brakes on a Toyota MR2.

Drum brakes are common on older muscle cars.

brakes often weigh less than original equipment disc brakes, and produce less friction when the brake pedal is not being pressed. The downside is that they often heat up faster if used hard.

Whether a brake system uses discs or drums, the brakes slow down the car by absorbing the energy of the car's forward motion and converting this energy into heat. The faster the car is moving or the heavier the car, the more heat the brakes must absorb. Too much heat build-up leads to trouble. If the brake pads or shoes overheat, they stop grabbing no matter how hard you push the pedal down. Having the brake fluid overheat will make the braking system go soft and may make it impossible to send enough pressure to the brakes to effectively stop the car. Many brake mods are designed to either help the brake system cool down or allow the brakes to still work when they are at a higher temperature. If your car has any problems with the brakes fading after repeated stops, it can benefit from a good set of brake upgrades.

The handbrake or parking brake is an exception to the rule about hydraulic brakes, except on some heavily modified drift cars. Instead of using fluid, the handbrake lever pulls on a cable to activate the rear brakes. While this usually operates the normal rear brakes, some cars with rear disc brakes have a miniature pair of drum brakes inside the center of the disc brakes to act as parking brakes.

Brake Rotors

There are several ways to make brake rotors either heat up more slowly or shed heat faster. The simplest way to make rotors heat up more slowly is to make them bigger. The more metal you have, the more energy you need to raise its temperature. If you placed a piece of very thin sheet metal on your stove and turned the burner to High, it would be glowing red very quickly. A thick cast-iron skillet, on the other hand, would take a longer time to get that hot. A larger rotor is the equivalent of the cast-iron skillet. Adding more rotor diameter will allow your brake rotor to absorb more heat and help keep temperatures under control. The downside is, of course, that these rotors are heavier.

The most effective change to a rotor to improve cooling is to add vents. A vented disc brake has a rotor with a set of cooling vanes separating the two surfaces in contact with the pads. The vanes act like an air pump, drawing air from the center of the rotor and sending it out the edge. This air flowing through the rotor carries away the heat generated by braking. If your car uses solid discs, converting to vented discs can mean a big difference in how well your brakes hold up.

Don't confuse vented discs with cross-drilled or slotted rotors. A vented disc can still have perfectly smooth surfaces, as the vents run down the center of the disc. Slotted rotors have a series of shallow grooves machined in the surface of the rotor. These grooves prevent the brake pads from becoming glazed—an effect caused by burned brake pad material sticking to the pad. This is a very useful feature on road race or circle track cars, which often heat their brakes until they glow bright red. The downside of

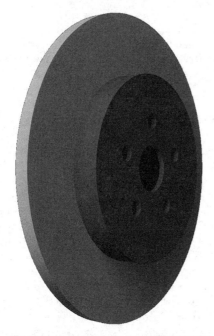

This very basic brake disc has room for improvement.

The vents, visible at the edge of this rotor, give it more cooling power.

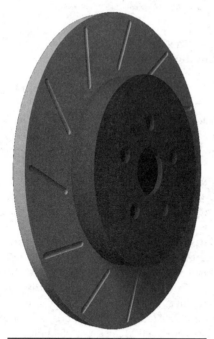

A slotted brake rotor.

A cross-drilled brake rotor.

slotted rotors is that they remove good pad material along with the bad. If your street car does not often suffer from glazed brakes, these may not be worth the wear.

Cross-drilled brake rotors were designed decades ago to deal with brake pads that released gases when they heated up. These gases would form a layer between the pad and the rotor, making the brakes ineffective. Drilling holes in the rotor let the gas escape into the holes. Cross-drilling quickly became established as "the way high-performance brakes look" in the public mind, a view that is so persistent that some sports cars come with cross-drilled rotors from the factory to this day simply to meet expectations. However, modern brake pads do not release gas like old pads did. Now, cross-drilled brake rotors only accomplish the same effect as slotting—they remove glazed material from the pad.

Cross-drilling brake rotors has one disadvantage that slotting usually doesn't. Holes in the rotor will create weak spots, particularly if the holes are drilled rather than cast into the rotor. Higher-end brake rotors often have smooth edges on the holes to reduce the odds of cracking; cheaper ones are often simply drilled with no attempt to smooth the edges, making the holes a fracture waiting to happen. If you want cross-drilled rotors, be sure to get high-quality ones from a reputable manufacturer.

There are two common ways to make a brake rotor. One is to make the entire rotor from a single piece of cast iron. The other way is to make the part of the rotor that contacts the pad from cast iron and bolt it to an aluminum center. These two-piece rotors are more expensive, but have several advantages over one-piece rotors. The most obvious is that the aluminum makes them lighter. A less obvious advantage is that the iron part is not firmly connected to the center. This gives the iron more freedom to

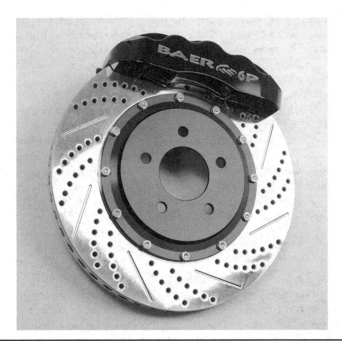

This rotor is vented, cross-drilled, and slotted at the same time. (*Photo courtesy Baer.*)

expand when it heats up, reducing the stress placed on the rotor by temperature changes. Less stress means the rotor can last longer without cracking. It is up to you to decide whether these advantages of a two-piece rotor justify the additional expense, if you can find them for your car.

If you feel a vibration in the brake pedal, it is usually due to the brake rotors. A common misconception is that this is from the rotors warping. This brings up a mental picture of the rotor becoming so hot that it turns into the shape of a Pringle's potato chip. In most cases, what actually happens is that the rotor manages to get some spots that are thicker than others. There are several ways that this can happen, most of which

One piece (left) and two piece (right) brake rotors. (*Photo courtesy Baer.*)

are due to heat. One cause is parking the car while the brakes are still hot from a few hard stops. This can cause part of the brake pad to melt onto the rotors, forming a thick spot. Heat can also cause small hot spots that wear differently. Moderate problems with uneven wear can be cured by having the rotors turned on a brake lathe, but severe problems can only be fixed by replacing both the pads and the rotors. One step you can take to reduce this problem is to cool down the brakes with slow driving and minimal braking for a while between any hard braking and the time you park the car.

If you are installing oversized brake rotors, you will probably have to replace some of the other parts. Most aftermarket big brake conversions will come as a complete kit, and may have replacement calipers or caliper mounting brackets as needed. If it is possible to install larger brakes off a different model car, you will need to make sure that you get all the junkyard parts this conversion requires. If you have small stock rims, however, these new brakes may not fit inside your wheels. Check with your supplier to make sure what sized wheels you will need to clear your brakes.

Brake Pads

Brake pads come in several varieties, with different materials meant for different uses. Of all the brake components, pads and shoes are the most critical to match to the way you drive your car. If you took every brake part out of a Nextel Cup car and installed it on a street car, the brake pads would be the least acceptable part for street use of any of the components. Likewise, if you are going road racing and can only afford one brake upgrade, make it the pads.

The cheapest sort are organic pads. These consist of wood and paper held together with a sort of glue called a binder. These pads do not stand up to high heat levels and do not have much "bite," meaning that you will need to hit the pedal harder with these than with other types of pads. Their biggest advantage, besides their low cost, is that a car that stays parked for long lengths of time is less likely to have a big patch of rust form under the brake pads.

Semimetallic brake pads are similar to organic ones, but contain a significant amount of metal as well. This gives the pads more bite and helps them last longer. Semimetallic pads are often a good choice for a street car if you want some improvement over original organic pads but would rather not spend what it takes to get the ultimate in braking performance.

The top performing brake pads short of the exotic hardware on Formula One cars use a compound that contains carbon, metal, binder, and little else. These pads often bite even harder than semimetallic pads and can withstand even higher temperatures. Brake pads like this are available from several aftermarket manufacturers, including Carbotech, EBC, Hawk, Porterfield, Performance Friction, and several others.

Even with the same sort of basic brake pad material, there are several ways the pads can differ. Some of them are designed for different temperature ranges. The higher the maximum temperature, the more the pads resist brake fade. Racing pads often do not work very well below a certain minimum temperature, and using them when they are too cold can mean less stopping power and more wear on your brakes. This is definitely a case where manufacturers really mean it when they say "For race use only." No one brake pad can fit every application from daily commuting to all-out road racing. Fortunately for the street enthusiast, a single brake pad can often work quite well for the combination of street driving, autocross, and drag racing. In all three cases, the best pad to use is one that gives good performance when cold and offers better fade resistance than stock. Pads meant for road racing or paved circle tracks will not work very well in any of these three uses. By the same token, good street brake pads would be a disaster on a road course.

There are several other important factors to consider when choosing brake pads. Some brake pads will suddenly grab once you put a certain amount of pressure on the pedal, while others are more predictable. A predictable amount of grab for how hard you push the pedal is known as modulation. Pads also differ in the amount of wear they inflict on the brake system and the amount of noise they make.

When just installed, brake pads should be "bedded in." This process applies a thin coat of brake material to the rotors and helps the pads wear into the shape of the rotors. Exactly how to do this will depend on the particular brake material you have chosen, but it often requires a series of hard slowdowns from 60 to 10 mph followed by a period of cooling off by driving the car slowly with minimal use of the brakes. Bedding in the brake pads will prevent premature break wear and help you get the most performance out of your new brake pads.

Calipers

A good set of aftermarket calipers can have their advantages. One is that they may fit that set of oversized aftermarket rotors you want to install, and your stock calipers don't. But other than that, there are several advantages an aftermarket caliper may have due to its construction.

These Wilwood brakes have six pistons in the caliper, three on each side.

Performance calipers are frequently designed for extra stiffness compared to the stock pieces. A stiffer caliper will require pushing the pedal less to get the same amount of clamping force. This can make for a noticeable improvement in the way your braking system feels. Many aftermarket calipers are also made from aluminum for lighter weight.

Calipers use pistons to push the pad against the wheel. Calipers may have anywhere between one and eight pistons, sometimes even more. The biggest advantage of extra pistons is that the extra size often allows for installing bigger brake pads. These bigger pads can absorb more heat and resist brake fade better. If you add more piston area to the brakes at one end of the car, the brakes at that end will grip harder compared to the brakes at the other end. You may need to change the calipers (or brake cylinders for drum brakes) at the other end of the car so that the brakes at the other end are doing their fair share of the work.

If you are considering calipers meant for race cars, be sure that the seals are designed to adequately keep out water. A seal that lets a little water into the brake lines every time it rains may not be much of an issue for a racing team that replaces the brake fluid after every race, but it is definitely a nuisance on the street.

Drum Brake Upgrades

If you have drums, there are still a few changes you can make to get more performance. While drum brakes may fade faster and behave in more of an on/off fashion than disc brakes, there are steps that you can take to minimize both of these. Since the front brakes handle the lion's share of the braking effort, a car with discs up front and drums in the back can still have decent stopping ability.

Just like with brake pads on discs, it is possible to upgrade the brake shoes, with much the same benefits. Even if you are unable to find off-the-shelf brake shoes, some companies can strip off the friction material from your original brake shoes and bind on new high-performance material. Be sure to select a brake shoe lining that works well with your front brake pads. Good modulation is even more critical for drum brake shoes than disc brake pads; you may even want to give up maximum bite to have more predictable behavior.

There are a few changes that can be made to the brake drums, too. In some cases, you may be able to find a set of wider or larger drums in a junkyard that can be swapped onto your car. Occasionally you may be able to find finned aluminum drums. These drums can dissipate heat much faster and look shiny while doing it. Sometimes all that needs to be swapped is the drum itself, while other times you will have to swap the hardware inside the drum as well. Keep in mind that larger brakes will also add weight.

In many cases, a manufacturer offers several different wheel cylinders that work with the same drum brake setup. Swapping these cylinders can be a useful tuning tool to adjust the brakes so that each end is doing its fair share of braking. If the front brakes lock up too early, a larger-diameter rear wheel cylinder can make the rear brakes grab harder. On the other hand, if your rear brakes lock up too soon, swapping to a smaller-diameter rear cylinder can cure this problem.

After making all these upgrades, however, you may still not be satisfied with how your larger brake drums work. If you know that upgraded rear drums will not give you enough stopping power for your tastes, or if you have drums up front, it may be time to consider a drum to disc brake swap.

Swapping Drums for Discs

Depending on what car you have, you may have several options for replacing drums with discs. Aftermarket kits may be the most convenient option, but they can also be the most costly. Sometimes your car may have a high-performance version sold alongside it that included discs, in which case you can have a relatively straightforward upgrade available in junkyards. If this is not an option, it may be possible to fit brakes from a later or larger car using a mixture of parts from several different junkyard cars. If you are using a mix-and-match approach, you will need good, reliable information from either a written guide that covers this swap or a fellow enthusiast who has already done this swap, unless you're willing to do a lot of careful measuring and calculations.

The amount of work involved varies. With many aftermarket kits, all you need to do is remove the drum, take off the plate behind it where all the drum hardware mounts, and bolt the parts from the kit on. Junkyard swaps may require disassembling half the suspension. As with upgrading disc brakes, the same warning about wheels applies. Some of these brakes will not fit under the original wheels. Some of the junkyard upgrades may not even have the same wheel bolt pattern.

Whether you are using aftermarket or junkyard parts, it is possible that you will have to replace or modify the master cylinder. Some drum brakes use a device called a residual pressure valve to hold a minimum amount of pressure in the brake lines to balance out the force from a spring that pulls the brake shoes together. You must not use this residual pressure valve with disc brakes. If you use this valve with a set of disc brakes, the pressure will keep the brakes slightly on at all times, slowing you down and heating up the brakes. This valve is usually located inside the master cylinder, but some cars have it mounted in the brake lines.

If you go with an aftermarket kit, check if it is intended for street use. Racing disc brake kits often have undesirable features for everyday driving. The same lack of seals mentioned on calipers can turn up here. Or a kit may leave out the parking brake. Disc brakes intended for drag racing may be designed only to save weight and have worse heat build-up problems than your stock drums. Brake kits need to be matched to how you plan to use them.

Brake Fluid

Your brake fluid needs to be capable of handling the heat from your brakes. Heat build-up can cause it to boil, reducing the amount of pressure you can send to your brakes. In extreme cases, the brake pedal may go all the way to the floor as though all the fluid has leaked out. Changing your brake fluid regularly is as important as running the right sort of fluid. Brake fluid tends to absorb water from the air, which lowers its boiling point considerably.

Most brake fluids contain mostly glycol. These include DOT (Department of Transportation) 3, 4, and 5.1. Ordinary DOT 3 brake fluid has a minimum boiling point of 401°F straight out of the can. This fluid is also the longest lasting. DOT 4 brake fluid raises this to 446°F, and DOT 5.1 has a minimum boiling point of 500°F. These all drop considerably when the fluid has absorbed too much water. Note that some heavy-duty fluids may have a boiling point above the DOT minimum rating. Racing brake fluids often far exceed DOT 5.1 requirements.

Make sure your brake fluid is able to stand up to the heat you encounter. (*Photo courtesy Edelbrock.*)

DOT 5 brake fluid, at least in the United States, means silicone-based brake fluid. This fluid does not absorb water. It also does not damage paint, which is a real problem with glycol-based brake fluids. Unfortunately, it is also somewhat spongy compared to glycol, and is seldom the first choice for performance use. Silicone fluid is often a good choice for show cars to protect the paint. If you are putting silicone brake fluid in your car, you must first remove all traces of the original brake fluid by flushing out the system with alcohol. Silicone and glycol must not mix.

Air in the brake lines is even more trouble than boiling brake fluid, since the air is in the lines even when the brakes are cold. Air can enter the brake lines through a leak or when you change the brake fluid. The calipers have bleeder screws you can use to get the air out, although it can be a time-consuming and messy process. There are a variety of products on the market designed to help you get the air out of your brake lines and make brake fluid changes easier. One product worth looking at is Speed Bleeders. These replace the bleeder screws at each brake, the plugs that allow you to drain the brake fluid. Speed Bleeders contain one-way valves, making it impossible for the brake system to suck up air when you are changing your brake fluid.

Brake Lines

The brake lines contain solid metal tubing or hard lines to carry brake fluid around the body of the car and flexible lines to carry brake fluid from the hard lines to the brakes at each wheel. There is not much you can do to improve the hard lines' performance unless yours have started leaking, although some shops offer stainless steel replacement

Speed Bleeders only cost a few dollars, and the one-way valve lets brake fluid and air out, without letting the air back in. (*Photo courtesy Edelbrock.*)

lines to better survive rust. You can make replacement brake lines at home with a double flaring tool, although you'll want to practice with it a bit on a few sections of line before making anything you'll put on the car. If making a replacement brake line, be sure to copy the loops and coils, which are there to help the brake lines survive vibration.

Ordinary brake hoses are made from rubber, which offers some room for improvement. Rubber hoses work quite adequately for normal driving, but their disadvantage is that pressure in the brake lines makes them swell up slightly. This does not rob your brake system of any grip, but it does require you to push down the pedal a bit further and makes the brakes feel somewhat rubbery. Stainless steel brake hoses

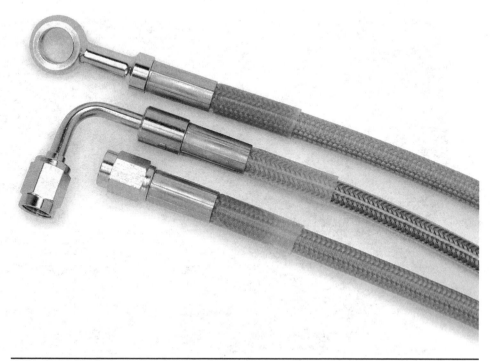

Braided metal brake hoses can be a good upgrade over ordinary rubber. (*Photo courtesy Edelbrock.*)

contain an inner plastic hose covered by braided metal. These hoses are not only stronger and more wear resistant, but the steel restricts how much they can expand under pressure. Replacing rubber brake hoses with stainless steel hoses can improve brake feel.

Proportioning Valves

The proportioning valve controls the ratio of pressure at the front brakes to pressure at the rear brakes. Cars generally come with a fixed proportioning valve that works very well with the original brake setup. The factory setup will probably behave quite well if you have made minor tweaks at one end or put a matching set of aftermarket pads on all four corners, too. Sometimes you can adjust bias by giving the rear brakes shoes or pads with a different material, or even swapping the wheel cylinders on drum brakes. However, major changes such as a monster disc kit or replacing drum brakes with discs can upset the original balance to the point that little changes to brake pads or wheel cylinders will not bring the car back to normal. If one end locks up before the other has reached its maximum stopping force, the brakes will no longer be as effective.

Aftermarket adjustable proportioning valves allow you to fine-tune your brake balance. Installing one requires removing the stock valve (running two proportioning valves at once can cause very unpredictable braking behavior) and mounting the new valve somewhere in the rear brake line. Race cars often have this valve located somewhere inside the car where the driver can easily reach it to compensate for

An adjustable proportioning valve lets you adjust the balance of your brake system.

changing track conditions. On a street car, mounting the proportioning valve somewhere under the hood is fine. This is still relatively accessible, and it cuts down on the risk of spilling brake fluid all over your carpet. Adjusting the proportioning valve to balance the front and rear braking power so the two ends lock up at the same time will give you maximum stopping power. For safety's sake, it's best to dial it in so the front brakes lock up just before the rear ones, because having only the rear brakes locked up can spin out a car. Since a badly adjusted proportioning valve can be unsafe, it is best to test your proportioning valve in a safe area where there is nothing to hit if you lose control.

Some cars use what is known as a split diagonal brake system. This uses two sets of brake lines, each controlling the wheels at opposite corners. The rear brake lines do not meet at any point in the brake system. If your car uses this sort of brake system, you cannot use a single proportioning valve meant for just one brake line. Either use a valve especially made for two lines, or install a separate valve in each line and be sure to set them to identical settings.

Line Locks

Tires often have more traction when warm. On the dragstrip, the best way to heat up the tires is to burn some rubber. Most front-wheel-drive cars can stay in place while doing this simply by holding the handbrake. Rear-wheel-drive cars don't have this option. You can hold down the brakes and floor it, but this puts a lot of wear on your rear brakes. Line locks solve this problem by allowing you to activate the front brakes on their own. These are mostly useful for drag cars and should not be used while the car is moving. Line locks also should not be installed on cars with antilock brakes or split diagonal brake systems.

Master Cylinders

The master cylinder may need to be upgraded if you have changed your brake calipers or swapped drum brakes for disc brakes. Installing calipers with more piston area can mean that you must install a master cylinder with a larger piston area to supply the new calipers with enough brake fluid. A larger diameter master cylinder will send more fluid to the brakes and require pushing the pedal less distance, but with more force. On the other hand, if your pedal travels a reasonable distance but requires stomping down too hard, you might need a smaller diameter master cylinder. If your calipers work well with your existing master cylinder, however, there is probably no need to upgrade here.

Some cars from 1966 and earlier use a single-reservoir master cylinder. This is a definite safety hazard; if the fluid leaks out at any point, you lose all your brakes. This can be especially alarming if you have just pulled out of a garage and started to back down a steep driveway, as I know from personal experience. If you have a system like this, switching to a dual-reservoir master cylinder, as used in modern cars, will keep you safer. These usually can get the brakes at one end sort of working, even if one half of the brakes have lost all their fluid. Upgrading will require the valves from a later dual-reservoir car along with an appropriate brake line setup. If you are getting the valves from a junkyard car, you can make notes of how its brake lines are set up, or even grab the entire brake line setup if they are in safe condition.

This 1966 Dodge Dart originally had a single-master cylinder, which failed while backing down a steep driveway. Replacing it with a double-master cylinder makes the system much safer.

ABS Controls

The antilock braking system is designed to back off the brakes when they are applied hard enough to lock up. This can help the driver maintain control under hard braking. Usually, there is not much that a tuner can do with an ABS controller. Some early ABS systems can hurt performance enough that it might be desirable to remove it for an all-out race car, but on a street car, it is usually best to leave the ABS system in place and working.

Cooling the Brakes

There's another way to deal with brake fade besides making the brakes absorb heat: carry the heat away by bringing fresh air to your brakes. A minimal approach is to use scoops under the front bumper that send air in the general direction of the brakes. For even more cooling, you can run a flex hose from an opening in the front or side of the car to the brakes. This hose should deliver its air directly to the caliper, which is likely to be the hottest point of the brake system.

Brake cooling is one area where you don't need to compromise. Short of dangerous or insane ideas involving CO_2 or liquids, it's hard to overcool your brakes.

CHAPTER 11

Interior

Safety, Function, and Style

Interior mods can do more for your car than just ensure you have a good-looking spot to sit in. Depending on your preferences, you may want to add gauges to monitor your engine, put in safety equipment, or simply go through your interior and remove any dead weight.

Rollover Protection

Some sedans may have a roof strong enough to support the car if you could set it gently upside down. Very few roofs are designed to take the blow of a car being flipped over violently. This is why racing frequently requires cars to use either a roll bar or a roll cage.

The interior of this Bonneville car leaves you no doubt it was built for racing. (*Photo courtesy DIYAutoTune.com.*)

A basic roll bar.

Before looking for rollover protection, check the rules for any events where you plan on racing. The rule book will list what sort of protection you need here. You may not need anything for autocross or a mildly modified drag racer. A nine-second drag car, top speed racer, or road race car is likely to need a setup that is approved by the sanctioning organization.

The simplest sort of rollover protection is a roll bar. This usually consists of four bars (and possibly a few diagonal braces) that set up a hoop behind the front seat. Roll bars typically add around 50 lb to your car's weight. They also can cost you a bit of street usefulness. Many roll bars make it nearly impossible to use the back seat. If you have a convertible, a race-legal roll bar may be so tall that it does not allow using the convertible top. The bar also adds an element of danger in a rear-end collision if you aren't wearing a helmet, since the impact may clunk your head against the bar. It is not as dangerous on the street as some other types of bars, but the added risk is there

Some companies offer "style bars," "monkey bars," or other assemblies designed to look like roll bars. These are not designed to take the impact of a collision at all. Many of these bars are downright dangerous. Some of them position the bars far too close to your head. Others might fold in a collision, leaving you to deal with a bunch of folded tubing when climbing out of a wreck. The scariest may be the ones that do not actually attach to anything, but simply sit on the floor. In a collision, this kind of "roll cage" is likely to move around, and being hit with a 50-lb piece of steel can be the difference between landing in the morgue and walking away from a crash. Leave monkey bars to the monkeys.

The next step up is a roll cage. Roll cages add bars running forward from the main hoop of the roll bar to prevent the front of the roof from collapsing. Roll cages are often

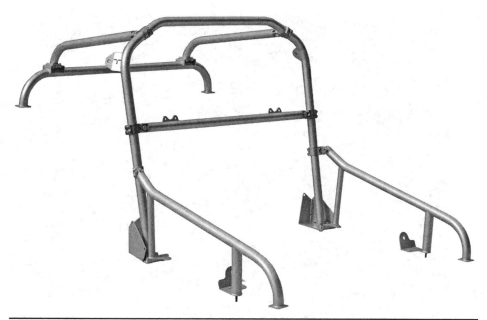

A more complicated bolt-in roll cage. (*Photo courtesy Ridetech.*)

described in terms of the number of points where they connect to the body of the car, with anywhere from six to ten or more points. The more points, the more bars and braces a roll cage will have. Some of them may require cutting the dashboard or running tubing through other parts of the car's interior.

A roll cage will clutter up your interior with tubing. This can make getting into and out of your car much more difficult. Most roll cages even include bars that go across the doors. If these bars are permanently welded in place, climbing into the driver's seat becomes quite a chore. Some cages include removable door bars to make this easier, but the trade-off is in safety. A door bar designed to be easily removed is more likely to get removed in a collision. For a street car that never sees any racing, a roll cage is often more trouble than it's worth.

There is another problem with installing a full roll cage in a street car. Roll cages are designed for a driver who is using the full set of protective gear required by racing organizers. That includes a helmet and a racing harness that keeps the driver strapped in far more rigidly than normal seat belts. Drivers wearing this gear will not have to worry very much about banging their heads against a roll cage. Unless you wear all that gear on the street, however, a roll cage can create a likely hazard (having the hard tubing crack your skull in a normal crash) while protecting against a much less likely one (rolling over is not a common problem on the street). If you aren't wearing a helmet, a roll cage can turn a survivable crash into a lethal head injury. Talk to enough racers and you will hear stories about someone who got killed by a roll cage in a street accident.

It is often a good idea to install padding on the roll bar, even though this may not protect you from banging your head against your roll cage with no helmet. This padding may slip on or attach with Velcro. Look for padding certified by SFI or other racing safety organizations, and for racing use, this padding should not have any upholstery

Land speed racing requires a massive roll cage if you plan to get your car above 175 mph. This 240SX is getting a custom roll cage built, and there are still some tubes left to install. (*Photo courtesy DIYAutoTune.com.*)

attached over it. Padding your roll cage with foam pool toys or pipe insulation isn't going to cut it.

Tubing for roll bars and roll cages comes in two sorts. Mild steel (sometimes called DOM, or *drawn over mandrel,* for the way it is made into tubing) is the less expensive choice, and is both widely available and easy for a shop to work with. Chromoly steel is a stronger alloy of steel. It allows you to use tubing with thinner walls and smaller diameters, but is considerably more expensive.

Roll cages and bars offer two choices when it comes to how to attach them. Bolt-in roll bars can be installed with just a drill, a wrench, and several hours of time. This is a mod that can be undone if you want to sell the car or decide you do not like dealing with all that tubing in your interior. Weld-in setups either weld to the frame for a full-frame car or weld to large metal plates that weld to the car's floor if you have a unibody. Usually, this is the stronger way if done correctly. If you are installing this for racing, be sure to check with the sanctioning body to see what sort of installation method they allow. Most rules require welded-in roll cages, but a few racing classes require bolt-in roll cages to discourage competitors from using the roll cage as a chassis brace.

Harnesses, Restraints, and Seat Belts

Ever wonder why NASCAR racers run at over 200 mph, but do not use air bags? One of the reasons is their highly effective seat belts. Racing harnesses are designed to hold

A five-point racing harness with cam-lock buckle. (*Photo courtesy Ridetech.*)

the driver much more firmly into the seat than a normal street seat belt. Not only will a properly designed and installed harness keep the driver safe on the race track, harnesses can also help the driver stay put during hard cornering. The downside is that these harnesses do not extend like normal seat belts. Once you adjust this, it stays rigidly in place, keeping you from leaning forward at all. A correctly adjusted race harness should be tight enough that it may take a bit of force to get it buckled.

Four-point harnesses resemble a normal seat belt with an extra shoulder belt. While they may look like a race piece, these have one serious flaw. In an accident, the shoulder belts tighten. Since the shoulder belts in a four-point harness attach near the center of the lap belt, tightening the shoulder belts pulls the lap belt upward. A driver can sometimes slide under the belt if this happens, an effect called *submarining*.

Most racing organizations require at least a five-point harness. This design adds yet another belt, called an anti-sub belt, that runs beneath the driver's leg. The anti-sub belt is meant to keep the lap belt firmly where it belongs rather than allowing it to slip upward in a collision. Six-point harnesses use two anti-sub belts for added security. Belts are available in several widths; naturally, the 3-inch-wide belts will be stronger than 2-inch-wide ones.

Besides the number of belts, there are a few other things to check for when buying a harness. The most important is to get an SFI- or FIA-certified one, whatever the rules require, since most racing events will not let an uncertified harness pass technical inspections. Harnesses also offer a choice of locking mechanisms. Standard latches resemble the buckles on airplane seat belts. They are cheap and rugged, but fastening and unfastening one can get a little complicated. Cam-lock harnesses have each belt snap into a single buckle that you can release with a small twist. This is much faster and more convenient to remove, but can get clogged if it gets dirty. This is not usually a problem on the street or in road races, but can be an issue if you race on dirt or salt.

Mounting a harness is as important as choosing the right harness. Attaching the shoulder straps of a racing harness to the floor is downright dangerous. The best place to attach a harness is the appropriate tube on the roll bar or roll cage. If you do not have a roll bar, the safest bet is to install a harness bar, a special bar that bolts to the car body and fits just behind the front seats. A harness bar is not as strong as a roll bar, but is a useful choice if you want to run a racing harness for autocross or similar events and also want to keep the use of the back seat.

The angle at which the harness points as it comes off your shoulder is extremely important. The best angle is somewhere between horizontal and angled down at 10 degrees. Angling up slightly is acceptable, but will not restrain you as well. If the shoulder belts angle down beyond 45 degrees, however, this can be extremely dangerous. The only way a harness with downward-pointing shoulder belts can hold you back is by pulling down on your shoulders. This can severely injure your spine. Check the directions that come with the harness to see what is a safe angle, and never, ever attach a harness directly to the floor behind your seat.

Harnesses and seat belts partly absorb the impact of a crash by stretching. Once the harness has been stretched, it is weaker and may break if you crash again. Always replace any harnesses or seat belts worn after a significant crash, and be very careful of buying a used racing harness unless you trust the seller not to sell you one that has already been in an accident. Most racing inspectors also require harnesses to be relatively new, since the belts can lose their strength over time. If your regular seat belts are looking faded or frayed, it is also a good idea to replace them. If a seat belt has taken enough damage from sunlight to break down its color, you can bet the sun has also damaged the material.

Racing harnesses are not the only driver restraints on a race car. Since many race cars run with the windows down, race cars need some means of keeping the driver's arms from sticking out the window, without making it too difficult for workers to pull anyone out of a wrecked car. On an enclosed car, this usually means window nets, like the ones used in NASCAR. Convertible drivers can use arm restraints, which are basically armbands on small straps that clip to a suitable point in the car's interior.

Some older cars only have lap belts. This is not really adequate for protection in a crash. Often, installing shoulder belts for the back seat may be straightforward, but it can be difficult to find any place to mount the front seat belts. The sturdiest solution would be to install a harness bar or roll bar and connect the shoulder belts there. Another option is to find a set of factory seats with built-in shoulder belts, like those in Chrysler Sebring convertibles, and adapt the seats to your car, making sure to use adequate sized bolts to attach the seats to the floorboards. One last option is to find a metal part of the car's structure and attach a plate there to mount the shoulder belt, preferably by welding the anchor in place. Just like with racing harnesses, it is not safe to simply bolt the shoulder belt to the floor.

If you have only stock seat belts but want the clamped-in-place feeling of a racing harness, you can add a CG-Lock to the belt. This device slips over the existing seat belt and locks out the extension mechanism with the flip of a switch. They are very handy to keep you from sliding around if you are into autocross. Shoulder pads are another popular mod for stock seat belts as well as racing harnesses. These are mostly cosmetic, but can make a tight racing harness feel a bit more comfortable.

Fire Extinguishers

Whether you want to build a race car, want to look like you have a race car, or are just plain scared of how the previous owner "fixed" the wiring with lamp cords and duct tape, you might want to carry a fire extinguisher in your car. However, simply grabbing the nearest one off the shelf at Wal-Mart can create more problems than it solves.

The first consideration when choosing a fire extinguisher is to make sure it can actually put out all of the sorts of fires that may happen in a car. Fire extinguishers are rated with an ABCD system. Class A fires include most burning solids, which can include your carpet or upholstery. Class B fires are burning liquids like gasoline. Class C refers to electrical fires. Class D means burning metals. Most cars can get by with an ABC fire extinguisher, although if you have magnesium parts on your engine, it's a good idea to keep a D-rated fire extinguisher in the garage (most D-rated extinguishers are huge industrial units, and not practical to carry around in a car). The "automotive" BC-rated fire extinguishers struggle to put out your seats if they catch fire in an accident or from a lit cigarette, making them a poor choice for keeping in your car.

In addition to making sure the fire extinguisher is up to putting out likely fires, it is vital to make sure the fire extinguisher is mounted correctly if you are carrying it in the passenger compartment. The plastic bracket that appears on many home extinguishers is not designed to hold up in a crash, and can send the fire extinguisher flying around

If you're working on cars, you will need a fire extinguisher rated for A, B, and C type fires.

the passenger compartment. The same is true with mounting your fire extinguisher to a piece of plastic trim. Screwing the fire extinguisher to the plastic trim on the windshield pillar is just asking for trouble. To be safe, get a metal mounting bracket for your extinguisher, and bolt it solidly to metal inside your car.

For racers and others who must have the ultimate in fire safety, the aftermarket offers more sophisticated fire suppression systems. A complete system contains a fire bottle connected to a set of metal lines and nozzles similar to a sprinkler system. Activating the bottle can instantly fill the engine compartment with Halon or other fire extinguishing chemicals without even having to open the hood. This can save precious seconds after a high-speed accident in a race car. Installing one for a street car is usually just an expensive exercise in paranoia.

Gauges

The right set of gauges can be a valuable tool for keeping an eye on what is going on under the hood. Unfortunately, many cars leave the factory without an adequate set of gauges. Some cars have only a speedometer and a fuel gauge. A few cars have been known to use "binary gauges," usually to indicate oil pressure. These "gauges" are actually controlled by an on/off switch, so the only readings are "acceptable" and "too low."

The aftermarket sells gauges to monitor a wide variety of functions, and these gauges sometimes come with a baffling variety of options. Some are simply issues of how the gauge looks, since it is easy for gauge manufacturers to change the color and materials they use. Others affect the way the gauge installs or operates.

Often, the original gauges aren't all you need to keep an eye on your engine.

The biggest difference in gauges is mechanical vs. electrical. Mechanical gauges are often the cheapest option for things that can be measured mechanically, such as temperature and pressure. Mechanical gauges almost always have a 270° sweep (the amount of angle the needle can turn), making them easy to read at a glance. The downside to mechanical gauges is that they need to connect to the engine compartment with thin plastic tubes. If one of these tubes is full of coolant or oil, having it break could produce a nasty spill in your car's interior. This breakage is rare, but if you are planning to put a mechanical fuel pressure gauge inside the passenger compartment, you must use a device called an isolator. The isolator mounts in the pressure tube, under the hood, and prevents any fuel from flowing into the line running into the passenger compartment. You definitely do not want to run any sort of tube full of gasoline through the passenger compartment.

Electrical gauges get their information from a device called a sending unit. The sending unit is a sensor that sends a signal back to the gauge. While some gauges may be able to work with your original sending units, most of them require buying an additional sending unit to work with your engine. Many of the less expensive electrical gauges have a needle with only a 90° sweep. The more expensive ones use a stepper motor to turn the needle, letting them have the same wide sweep of mechanical gauges. Some catalogs may call these "full sweep" gauges.

Digital gauges have a distinctive look to them. Many of them allow you to read them more precisely than a normal gauge. The downside to digital gauges is that they are not easy to read with a quick glance. Most drivers find it much faster to glance at a set of conventional gauges and notice that the needle is pointing in the safe zone than to look at a digital gauge, read its numbers, and determine if anything is wrong.

Some gauges are liquid filled. This liquid is designed to prevent the needle from fluttering due to vibrations or pulsing signals, making them easier to read. Liquid-filled gauges have a bubble visible at the top. This bubble lets the liquid expand on hot days, and serves to let everyone know you have spent the extra money on your instruments.

Tachometers often come with many unique options seldom found on other gauges, since the engine RPM is often the most critical factor for a driver to know during a race. The most common is the shift light, a very bright light triggered when the tachometer hits an RPM that you have chosen. Some of the more expensive tachometers feature a playback mode where they can record your RPM and let you know just when you shifted in the race. Also note that most tachometers are designed to work only with engines that have four, six, or eight cylinders. If you have an unusual engine with a different number and want an aftermarket tachometer, be certain that it can work with your motor.

Besides the set of a speedometer, tachometer, and fuel gauge, there are many gauges an enthusiast can chose from, depending on what features the car has. Many enthusiasts consider gauges for water temperature, oil pressure, and either an ammeter or voltmeter to be almost essential, while other gauges can be useful for certain situations. Here are some of the most common diagnostic gauge types:

- **Coolant or water temperature.** Overheating can do a considerable amount of damage to an engine, and can often be spotted before things are too late. Many enthusiasts consider this to be the most important gauge for avoiding engine damage.

- **Oil pressure.** Often considered the second most important gauge for determining the health of an engine. The oil pressure is typically low at idle, moving up as the RPM increases and down as the engine slows again. Low oil pressure can indicate wear, while having the pressure drop under cornering, acceleration, or braking can be a sign you need to baffle the oil pan.

- **Oil temperature.** Some enthusiasts like to know this temperature, although it is usually not as important as water temperature. If you have an air-cooled engine like an old Volkswagen or Porsche, however, this may be the only liquid temperature you can measure.

- **Ammeter.** Measures the flow of electricity from the charging system.

- **Voltmeter.** Measures the overall voltage level in the electrical system. Either a voltmeter or an ammeter can provide you with a good picture of whether the electrical system is operating properly.

- **Vacuum.** A vacuum gauge can be useful in troubleshooting an engine. Trying to drive with as much vacuum as possible can also improve your gas mileage. However, it is not particularly useful in a racing situation if the engine is working correctly. Many tuners prefer to use a temporary one for repair purposes rather than add an extra gauge to keep an eye on.

- **Boost.** This is very useful on a turbocharged car, and may be of interest if you have installed a supercharger. Nearly all aftermarket boost gauges also work as a vacuum gauge when the engine is not under boost. Most original equipment boost gauges are not known for their accuracy—even when they actually have their units marked.

- **Nitrous pressure.** Measuring the pressure in nitrous oxide lines is very important if you are running the bottle.

- **Fuel pressure.** Many enthusiasts find this mostly useful for tuning purposes, so they are often installed under the hood rather than where they can be seen while driving. These are very important if you are running nitrous oxide, however. If you have the nitrous on and the fuel pressure takes an unexpected drop, switch off the nitrous immediately.

- **Air/fuel ratio.** The less expensive ones use the original oxygen sensor and really only will warn you if the engine is very far into the dangerously lean zone if you are using a turbocharged or supercharged engine. More expensive ones use a wideband oxygen sensor and can be used to tune your engine for maximum performance.

- **Exhaust gas temperature.** This gauge is another useful tuning tool. Higher temperatures usually mean the air/fuel ratio is leaner. If the temperature climbs too high, it is definitely cause for alarm. However, "too high" can depend on the engine.

- **Brake pressure.** Some people like to use these to tune brakes, but this is often a distraction when you are not tuning. If you want to use these gauges, it is best to install two, one for the front brakes and one for the rear.

- **Transmission temperature.** This one is most useful if you are towing in a vehicle with an automatic transmission.

Gauges mounted in an aftermarket A-pillar pod.

Sometimes it is better to have an alarm than trying to decipher a gauge. This is especially true in autocross or other events that require quick decisions. The aftermarket offers quite a few warning lights with adjustable trigger points if you do not want to be distracted by looking at your gauges.

After choosing your gauges, the question is where to put them. One traditional, if crude, way to install a large tachometer is simply to fasten it to the steering column with an oversized hose clamp. Recently, A-pillar mounts have become popular, letting you attach the gauges to the windshield frame where they are easy to see at a glance. Gauges can be put into holes drilled in the dash or mounted in bolt-on pods above or below the dashboard, too. When mounting the gauge, be sure to consider whether it is in a location that you can quickly look at without taking your eyes off the road. One other popular trick for quick gauge readings is to rotate all the gauges so that the needles point straight up and down when everything is working correctly.

Seats and Upholstery

When it comes to seats, there are plenty of options. If you do not like the seats you have, you can replace them with seats out of a different model, get a set of aftermarket seats, or reupholster the ones you have. Which choice you want depends on what you need and what you can afford.

An aftermarket racing seat.

Factory seats have several good points. Many of them are designed to be comfortable for a wide variety of drivers. Seats from well-optioned cars offer a wide range of adjustment and power options. Some offer side-impact air bags, and all modern factory seats have had to pass a crash test. Using stock seats can often be the best choice for a car that sees primarily street use.

However, you may find that some other car's stock seats are a much better choice than the ones that came with your car. This is particularly true if you have an older car. There are a couple of issues to consider if you want to swap seats. First, measure the seat to be sure it is not too wide for your car body. Second, check the seat belts. Some cars use seat belts that partially mount to the seat, while others have all the points attached to the car body. If your car has seat belts that attach to the seat at any point, make sure it is possible to move the hardware from your old seats to the new one, or find a secure point on the floor where you can mount an aftermarket seatbelt. Third, swapping seats with power options can be tricky if the controls aren't on the seat, and trying to make side-impact air bags work in a car not originally equipped with them is nearly impossible for an amateur. While seats are not likely to line up with the mounting holes, this can often be fixed with a drill and sometimes a few spacers (which, in many cases, can be a stack of washers). Be sure to use bolts at least as large as the ones that held in the original seat.

Stock seats do have their drawbacks, though. They are often somewhat heavy. Many of them do not do a good job of keeping the driver in place during hard cornering, although some cars come with fairly heavily bolstered seats from the factory. In some cases, racing rules may insist on a seat with a fixed back, because having a seat back fold up in a collision can cause serious injuries. Many aftermarket seats are also designed to work well with a five-point racing harness.

Aftermarket seats may be designed for street use, racing, or both. Check the features carefully. Some aftermarket seats recline like normal bucket seats, while others are solid and have no adjustment whatsoever once bolted into the car. Many can be ordered with mounting kits designed especially for your car. Many of these seats, particularly rigid ones, are designed for light weight.

Pay careful attention to the strength of a seat. An all-plastic racing seat is not likely to hold up well if something hits your car from the side, and probably would not be allowed in any sort of racing besides drag racing. One good sign that a seat has been designed well is certification. Although SFI only certifies a few specific types of seats, the French FIA and German TÜV have strict criteria for certification for a wider variety of seats. Although there are some good seats out there whose manufacturers have not bothered with certification, having a seat certified gives you independent evidence of its safety.

If at all possible, see if you can sit in a seat before buying it. A seat that is comfortable for one person may be downright painful for another. This is especially true of racing seats, which often have large bolsters to keep you in place. Bolsters that are too wide may not be very helpful, while bolsters that are too narrow can make you feel as if you are being squashed in a vice.

Upholstery offers many opportunities for customization. Common options for seat covers include the one-size-fits-all covers in parts stores, but you can also order covers meant especially for your seats from a catalog, or have a local upholstery shop make new covers for your seats.

The off-the-shelf covers frequently offer an inexpensive way to protect your upholstery or give your interior a new look. Unfortunately, their universal design often means they do not fit especially well. These are typically made of a stretchy fabric that can work acceptably well if your seats do not have any large bolsters. They do not work particularly well for heavily bolstered sports seats. They also don't belong on seats with side-impact air bags. But they can still be a cheap way to spruce up your interior if your seats are starting to wear through.

Mail-order custom covers can fit a greater variety of seats. They may require a bit more work to install than a stretchy cover, but the result can often look close to factory installed. Many of them are attached to the framework of the seat with hog rings—the exact same sort of ring that farmers stick in pigs' noses. While some catalogs offer specialized hog ring pliers, installing hog rings with normal pliers is not too difficult.

For maximum quality or a truly unique look, your best choice may be to take your seats to an upholstery shop. There are few limits to what sorts of covers an upholstery shop can make, and they can also work with headliners, carpets, and other fabric interior trim. One very common trend in muscle cars is to take modern front seats and recover them with fabric that matches the original seat covers, combining classic looks with modern comfort.

Race cars frequently have all their upholstery removed except for the seat padding. Not only does this save weight, but it cuts down on the risk of fire. If you are removing the upholstery from an older car, think twice before throwing away any upholstery that is still in serviceable condition. You may have a hard time finding good replacements if you change your mind.

Trim and Color

A stock interior can be like a blank canvas, and every inch of it can be painted or retouched. Sometimes the aftermarket may offer ready-made solutions, while at other times you may need a little creativity to get the look you want.

Replacing controls like shift knobs, pedals, and steering wheels is very popular. Adapter kits allow the aftermarket to make wheels and shift knobs that fit a very wide range of cars, sometimes using adapters that come with the part and sometimes using adapters sold separately. Most aftermarket pedals simply clamp on to the original parts, although some, such as the Lokar street rod pedals, are complete assemblies that can require minor fabrication to install. There are a few things to watch out for when installing these. Some aftermarket shift knobs may not work if your shifter has any sort of knob you must pull to shift into reverse. Air bag steering wheels are another possible problem. It is usually not legal to remove air bags from a car, nor are air bags particularly safe to work with. The "air" in the air bags is supplied by a small explosive charge that can detonate under the wrong circumstances.

Some of the more popular modern cars have a ready supply of replacement trim pieces as well. Plastic trim inserts can be replaced with wood, carbon fiber, or metal, depending on the personality you want to give your car. The original gauge faces can often be replaced with pre-printed plastic to change the color or add graphics to the gauges. If you cannot find aftermarket gauge inserts but have a talent for computer graphics, some local sign shops can make custom gauge inserts to match your design.

If you cannot replace the trim, you may opt for simply changing its color. Some companies offer pre-cut vinyl decals to cover the trim with fake materials, but these do not always look convincing—especially if they start to peel. Painting the trim may not make it look like anything but painted plastic, but it can be a very economical way to give your interior a new look. Some paint companies offer special spray paints designed just for use on plastic.

CHAPTER 12

Exterior

Aerodynamics, Lighting, and Appearance

Just like with engine and suspension mods, exterior mods are most effective when they are designed to work together. This can be true of either functional mods or fashion. Automotive fashions come and go, but some principles—both aerodynamic and aesthetic—are fairly timeless.

From an aesthetic standpoint, the first consideration is the overall lines of your car. Most cars are built around a certain type of shape. A Honda CRX, for example, is made of mostly straight lines and nearly flat surfaces. The third-generation Mazda RX-7 has a shape dominated by curves, with very few straight edges. A body kit that looks perfect on one of these cars will look absolutely wrong on the other, because the overall shape is so different.

Stay true to the lines of your car, and carefully chose which features and lines you want to highlight. Usually, the lines you may want to highlight are things like styling creases in the sheet metal, while gaps between body panels are best ignored. Horizontal

Styling trends come and go, but some principles don't change. (*Photo courtesy DIYAutoTune.com.*)

lines often make a car look longer and lower. Vertical lines, such as adding chrome to the edges of your doors, often make a car look taller—not usually a desirable effect. Also, take care with how many lines you add. Too many lines can make a car look cluttered and busy.

If your car has been well used or damaged, the first step toward making it look good is to repair it. If you combine body mods with bad paint, rust, or dents, the flaws will often be what people notice first. Sometimes you may install the mods as part of the repairs or want to wait until your whole body kit is in place before painting the car. This can often make for a better finished product if the whole car is painted at once. A work in progress is understandable. Trying to pretty up a car with a wing when it needs major body repairs, on the other hand, can make a car look trashy.

Another important consideration is the purpose of the car. Cars often look best with appearance mods built around a single concept, usually a particular era or racing event. A well-chosen theme should match the strengths and styling of the car. An aggressive front airdam, splitter, aluminum wing, and other road racing parts would look very out of place on a PT Cruiser. Normally, the old-fashioned styling of a PT Cruiser would make street rodding–style touches a more logical choice. Likewise, an '86 Corolla GTS would be a very unlikely choice to equip with Cragar S/S wheels, scalloped paint, and other '60s touches, but would go well with a theme inspired by Japanese drift racing events.

Racing is one of the most popular sources of inspiration for customizing. Even many nonfunctional cosmetic parts are patterned after functional parts from one sort of race car or another. Different sorts of racing call for different aerodynamic aids, ride heights, and wheel and tire packages. Mixing and matching cues from different motorsports can often leave a car looking like it lacks purpose.

Drag Racing

Drag cars are designed around traction and power. The most important feature for a drag car is large enough tires on the drive wheels to put the power to the ground. Some rear-wheel-drive cars use tubbing, a mod to the rear floor pan to get more room to fit oversized tires under the car. The mammoth driving tires are often paired with small, lightweight wheels and tires at the opposite end, for a package known as *big 'n' littles*.

Drag car stances have varied over the years. The "Gasser" look of the '50s and '60s called for a lifted suspension at both ends, often replacing the front suspension with a solid axle for light weight. This then gave way to a raked stance, with the front end lower than the rear. At first, drag racers left the front end at its original height and raised the tail end. Heights gradually came down, with many modern drag cars running a lower-than-stock front end and the rear set at or below the original ride height.

Extremely high-powered rear-wheel-drive cars can benefit from a very large but nearly flat rear spoiler. Less powerful drag cars often lack wings or large spoilers, since unless you make enough power to need a wing, the extra drag can slow you down. Sometimes, however, a small wing can be useful to keep the rear tires planted firmly on the ground when you hit the brakes at the end of the run.

Although most drag cars use few cosmetic mods, one popular area to modify is the hood. Many drag cars use hood scoops to draw in cold air, and sometimes to let a tall intake setup fit under the hood. Both forward-facing and rearward-facing (cowl induction) hood scoops are popular among drag racers. Some racers who use

This Nova exemplifies the typical drag car look.

superchargers, tunnel rams, or other tall intake manifolds prefer not to use a hood that clears their intake system, and instead cut a hole in the hood to show off their engine to the world.

Most of the other visible mods on drag cars tend to concentrate on safety. Two of the most visible safety mods are parachute brakes and wheelie bars, both of which mount to the rear of the car.

Road Racing

Road racers often use considerably more aerodynamic mods than drag racers if the rules allow. While drag cars need most of their traction only at low speeds, a road racer needs full grip all the time for cornering. Preventing lift and creating downforce are critical. Road race cars often use airdams, splitters, spoilers, ground effects, and small-to-medium sized wings to keep the tires planted firmly at both ends. Many popular street body kits over the years have taken their cues from road race cars, from the wide-fendered look of the late '70s to the '90s-era Veilside kits.

Large, low-profile tires at both ends are also important to getting good traction. Many road race cars use equal-sized tires on all four corners, although some extremely powerful rear-wheel-drive cars may use larger rear tires. Tubbing is not very popular with road racers, who often prefer to fit larger tires by pushing the outer fenders outward with flares or wide-body kits. Most cars built to handle have a lowered stance, unless you are dealing with a car that has suspension geometry problems when you lower it.

The Condor Speed Shop BMW is set up for road racing, with flared fenders and big tires all around. (*Photo courtesy DIYAutoTune.com.*)

Hood scoops are less prevalent in road racing than drag racing. However, road racing cars often need scoops and inlets to cool brakes and the engine compartment. Cooling is a major issue, particularly since some racing events can last 24 hours straight. Some cars use extractors on the hood designed to vent hot air from the engine compartment, which also cuts down on lift in the bargain.

Autocross cars often use the same tires and stance as a road race car, but the aerodynamic aids are somewhat less common for chasing cones. This is partly due to rules that often ban wings and spoilers, and partly because most wings do not work very well at autocross speeds. A few autocross cars in anything-goes classes like A/Modified run wings that are close to a third of the size of the car, but these are usually not based on production cars at all.

Drift cars also tend to copy the road racing look, with a Japanese flavor added. They often use the same sorts of wings and body kits, but frequently also take inspiration from Japanese car customizers. One other common touch for drift cars is that the graphics on each side are often mirror images of each other, with the text on one side reversed. It remains to be seen if the Japanese influence on drift cars will make it onto the growing number of American cars being entered in drift events, as American drivers have begun experimenting with drifting Mustangs, Camaros, and even El Caminos.

Top Speed Racing

Cars built for maximum speed need minimum drag, if the rules allow it. Some classes don't allow body mods at all, but lowering the car and changing the wheels are fair game. One very distinctive touch on top speed cars is the nearly flat aluminum wheel covers to reduce the drag created by the wheels, usually combined with skinny tires for less rolling friction. Classes that allow more body mods start with letting in airdams or

Black Opel Racing's Opel GT after running over 200 mph at Bonneville. (Photo courtesy DIYAutoTune.com.)

scoops and cowl hoods, much like drag cars. More extreme classes allow lowering the roofline or extending the nose. The top "special construction" classes were never production cars to begin with, and often look like Buck Rogers rockets on wheels.

Rally Racing

Rally cars are designed around the hazards of racing on dirt and gravel. While downforce is still important, low suspensions, airdams, and low-profile tires are not likely to stand up to repeated bumps or getting the car airborne. Consequently, many rally cars run at the stock height with somewhat smaller wheels and taller tire sidewalls (not to mention knobbier tire patterns) than road race cars. Rally cars often use wings like road racers, although recently, special rally wings have appeared designed to keep producing downforce while sliding sideways.

Many of the other mods on rally cars are designed to deal with rally racing hazards. Rally cars frequently run large clusters of extra lights to deal with racing on dark roads with no street lights. Mesh protects intercoolers and radiators from debris. The underside of rally cars is protected by skid plates and heavy-duty mud guards designed to keep the gravel from damaging the undercarriage. Since rally races are run with the windows up, many rally cars also feature a scoop mounted on the roof to draw in fresh air for the driver and navigator.

Importing Inspiration

Besides racing, one popular source of inspiration is overseas. Many cars were sold with slight (or sometimes even major) cosmetic differences in different markets. Some were even sold under different names, such as the car sold as the 240SX in the United States being related to the 180SX and Sylvia from the Japanese market. Owners of Japanese cars often look to JDM (Japanese domestic market) cars for sources of customizing parts, while European car owners can also get different parts from their car's homeland. Even owners of all-American cars can sometimes import different-looking parts if they know where to look. For example, the Dodge Dart had versions sold in South America that looked subtly different from their U.S. market cousins. The international nature of the Internet often makes it a good place to look for pictures of how cars differ in different areas of the world.

Customizing your car with imported factory parts can often make a very subtle statement. Most of the parts you are likely to find are unlikely to scream "Hey, look, this is a mod!" to the world. People unfamiliar with your car may realize, "Hey, that looks a little different." Those in the know will realize what you've done. If you want to set yourself apart from the cars people see every day while avoiding getting a ticket from the fashion police for excessive use of bling, this could be the approach you're looking for. Those who prefer a more blatant approach may still find foreign-market sheet metal useful for helping to install body kits intended for other markets.

A word of caution is in order; sometimes cars with the same name may be entirely different for different markets. For example, for many years Honda sold completely different Accords in the United States and Europe. Parts from these cars do not interchange at all. At other times, the years might not match, with car body styles being introduced and discontinued at different dates in different parts of the world.

Mod Materials

Body mods come in a wide variety of materials. These all have different strengths and weaknesses. Some are lighter, some are tougher, and some are easier to customize.

Metals

Steel may be heavy, but it is also tough and easily repaired. About the only thing it does not hold up well against is road salt. The most common off-the-shelf steel parts used in customizing are imported parts installed as original equipment in other countries, but some companies offer aftermarket steel hoods with a variety of scoops and other shapes.

Steel parts can also be modified. If you want louvers, a local machine shop can punch louvers into your original body panels. More complicated features can be made by welding scoops or other shapes to the original metalwork. When done correctly, a welded-on feature becomes practically a part of the original panel instead of reducing the strength of the part. Making new steel parts completely from scratch can be tricky if you need them to curve in more than one direction, but is possible for a well-equipped shop.

In addition to its toughness, steel has a certain prestige among street rod owners, and to a lesser extent, muscle car owners. This is partly because steel parts are usually either original or fairly expensive parts compared to fiberglass, and partly because steel simply has a feel of solidness to it.

Aluminum parts are somewhat rarer. Other than aluminum wings and the rare original equipment body panels used on cars such as the Acura NSX and Plymouth Prowler, most aluminum panels are hand built by careful cutting, hammering, and welding. The result is a metal panel that is very lightweight and very expensive. Hand-built aluminum panels tend to be somewhat delicate, too. These were used on older racing cars before the widespread use of lightweight fiberglass and carbon fiber. Few custom or race cars today use hand-built aluminum other than top-of-the-line street rods and cars that started out that way, but this can be an option if you want metal and have money to spare.

Many custom grills are made of simple metal mesh. Although you can buy mesh from some automotive catalogs, keep in mind that industrial supply houses such as McMaster-Carr also sell mesh and perforated metal, often at a lower price and with more varieties. Mesh is available in aluminum and stainless steel for durability. Do-it-yourself grills are often secured with carefully hidden zip ties if there are no convenient places where it can be attached with screws.

Composites

A composite material consists of fibers (usually made from ceramics such as glass) embedded in plastic or epoxy. The result is much stronger than plastic alone and much more resilient than a ceramic. The fibers are usually woven into a fabric, but do not have to be. Composite materials can make for very lightweight body panels that are sturdy enough to stand up to street use, although they sometimes need a few reinforcements that metal does not require. For example, most composite hoods need hood pins to keep them safely in place.

Composite parts are molded to their finished shape. Although there are shops that can design a composite part to order, a composite part cannot be modified like a metal one once it is complete. While a body shop can make small tweaks to fill in holes, anything that requires bending or forming is not an option. If you wish to add a scoop to a fiberglass hood, your only option is to cut a hole and attach the scoop with glue or double-sided tape.

Fiberglass is the cheapest and most common composite material. There are several types of fiberglass. Some of the cheapest fiberglass parts use small threads of unwoven glass fibers, which often results in a part that is heavier than an equally strong metal part. Woven fiberglass is more common for car parts, and it comes in two grades: the more common E-glass and the stronger but more expensive S-glass. Fiberglass parts are usually made from E-glass if they do not say otherwise.

Fiberglass is not as rigid as steel, and is considerably more brittle. A fiberglass body panel may visibly shake from engine vibrations if it is too thin, and fiberglass is likely to crack if anything strikes it. Many fiberglass part suppliers offer body panels in two versions: street and race. The street versions are thicker and heavier to make them rigid enough to give a solid appearance, although they are still likely to crack if they are hit.

Carbon fiber is definitely the trendiest composite at the moment. Pound for pound, carbon fiber is considerably stronger than steel. It is also much stiffer than fiberglass, although it is just as brittle. This is the material used to build Formula One race cars. Some manufacturers will refer to carbon fiber as GFRP, which stands for graphite fiber reinforced plastic.

Carbon fiber also has a distinctive look that is nearly impossible to imitate—not that this has stopped some appearance mod manufacturers from trying—and the results are usually not convincing. Unpainted carbon fiber has a dark, shimmering, high-tech look to it that is often prized as much as its light weight. Unfortunately, the epoxy used with carbon fiber deteriorates slowly when exposed to sunlight, meaning a show-quality carbon fiber hood may not remain show quality if the car is ever parked outdoors for years on end.

One other important composite is aramid, better known by its trade name, Kevlar. Kevlar is far less brittle than other composites, which is why it is used for body armor.

Much of this Eclipse's bodywork has been replaced with unpainted carbon fiber. (*Photo courtesy DIYAutoTune.com.*)

An aramid body panel will be far more impact resistant than one made from fiberglass or carbon fiber. Kevlar should always be painted, as it is quite vulnerable to sunlight.

The different sorts of composite materials can be combined using layers of different fibers. Many carbon fiber race car bodies also contain a layer of Kevlar for impact resistance. Kevlar is sometimes also combined with fiberglass for much the same reason. Those who want the look of carbon fiber at a lower price may want to have a composite made with an upper layer of carbon fiber but with fiberglass on the inside. It is even possible to weave fabrics containing several different sorts of fibers into one fabric. There are very few limits on the mixing and matching possibilities.

Plastics

When Saturn cars first came out, one of their selling points was the plastic bodywork. One Saturn commercial showed a car refusing to be dented even after being pummeled with golf balls. This is the biggest selling point for plastic aftermarket parts, too: They do not dent easily like metal, nor are they likely to crack like composites. Plastics come in a bewildering variety (Saturn used four different blends of plastic to make different panels on just one car), but most aftermarket plastic body panels are either polyurethane or acrylonitrile-butadiene-styrene (ABS) plastic.

Plastic's dent resistance comes from its flexibility. The downside to this flexibility is that plastic parts will bend under heavy loads, including air pressure. Plastic parts are best used for areas that can be supported from behind, unless the designer has put careful attention into shaping the part for strength. Hoods and wings are particularly poor choices for all-plastic parts, since they are likely to sag under the force of the wind. There's a reason Saturn went with old-fashioned steel for the hood.

Polyurethane (sometimes simply called urethane) is an extremely rubbery plastic. Many original equipment bumpers are made from polyurethane, as are suspension bushings. Its extreme flexibility makes it very well suited to this task, as well as for airdams and other parts that may be struck by rocks or banged against the ground if the car's suspension bottoms out. Polyurethane bumpers should be supported using the original bumper supports so that they keep their form. Some drag racers have found that their bumpers can fold in and create more drag at high speeds if they remove the supports, offsetting the gains made by reducing weight.

ABS plastic is a popular material for hood scoops and some other bodywork add-ons. It is also the material Saturn chose to cover their doors. This is a much more rigid plastic than polyurethane, so an ABS hood scoop can stay together under conditions where a urethane hood scoop would visibly bend. ABS plastic scoops can be riveted to the hood, but are often held in place with double-sided tape made specifically for attaching body panels.

A few customizers have experimented with hoods made of transparent plastic such as acrylic (Plexiglas) or polycarbonate (Lexan). Transparent hoods first appeared on '50s-era hot rods and show cars, but have recently started appearing on a few imports. This can make for an interesting display on a show car. Unfortunately, plastic hoods cannot take heat like metal or composites. Acrylic begins to get soft around 155°F, meaning that the heat coming off your headers can easily start to melt a Plexiglas hood. Polycarbonate is more temperature resistant, but also more expensive. These transparent hoods also require considerable maintenance if they become scratched.

Painting plastic requires the correct chemicals. Some plastics need special primer coats. Polyurethane requires additives to make the paint match its own flexibility, since brittle paint is likely to flake off if the plastic underneath it flexes.

Aerodynamic Mods

The air flowing over a car creates two problems when it comes to performance: lift and drag. Drag is a force that pushes backward on the car and makes it need more force to move forward, while lift pulls the car upward so the tires have less traction. Drag occurs because the car must push the air out of its way, and some shapes push less gently than others. Lift is caused by the difference in air pressure between the upper and lower surfaces of the car. The air must flow faster over the top of the car due to its curved shape. Air pressure drops as it moves faster, so the car has more pressure under it than above it, pushing it up. Lift is virtually always an undesirable effect, reducing how much grip your tires have.

The opposite of lift is downforce. Downforce can be generated through a variety of means, and helps to keep the tires planted firmly on the ground. Unfortunately, many means of generating downforce also create additional drag. Downforce also needs to be balanced; a car with more of its downforce at one end is likely to lose traction at the other if pushed to its limits.

The air flowing over a car body does not always stay with the bodywork. In places, particularly the rear of the car, the airflow separates off the body panels like a waterfall coming off a rock, leaving a pocket of slow-moving, low-pressure air behind the car.

Air flows smoothly down the tail end of fastback cars like this Corvette.

The shape of the parts in this pocket has very little effect, and if it does, it may not be the same effect as these parts would have in a free airflow. As a general rule, if part of the back of a car slopes down within 34 degrees of vertical, the airflow will separate from it once the car reaches highway speeds.

Most of the separation occurs at the rear of the car. Different designs create separation at different points. Most ways to design the back of a car fall into three different categories: sedans, coupes, or "notchbacks" with distinct trunks; fastbacks; and blunt tails. Some cars have been available with all three designs on one model, such as the '64–'66 Plymouth Valiant/Barracuda and the '88–'91 Honda Civic/CRX.

Fastbacks often create the least drag, with air flowing smoothly down the rear window and only separating at the very end of the car. Sedans may have the flow separate from the rear window, depending upon how steep it is, run across the trunk, and then separate again at the end of the trunk lid. Blunt tails often produce the most drag, with the airflow separating above the rear window. Many blunt-tailed cars drag a horizontal vortex behind them, causing the airflow to run up the back of the car and circulate around.

Of course, the different body styles also often have different weights, varying weight distribution, different amounts of chassis rigidity, and other factors to consider besides aerodynamics. Some cars even use different wheelbases and overall lengths depending on the body style. One common example of a car with different body styles is the 5.0 Mustang, which came in fastback and notchback form. Most racers favor the notchback for its light weight and stiffer construction, even though it produces more drag.

Many body mods are designed to improve aerodynamics—or at least look like parts that do this. The best manufacturers will test their body kits and the like in a wind tunnel. Others imitate these kits, go by guesswork, or simply design something they think looks good. While it can sometimes be difficult to tell which design works the best without wind tunnel testing, a well-informed enthusiast can often tell which parts are likely to be functional and which are likely to be fashion accessories.

Wings

Automotive wings resemble upside-down airplane wings, and are designed to create downforce on the rear tires. They can create downforce by two methods: by directing the air flowing over them upward, and by using a curved lower surface to create an area of low pressure under the wing. To be effective, a wing should be mounted high enough to be in an area of smooth airflow. To keep the airflow over the wing from interfering with the airflow over the body, the wing supports should be at least as high as the length of the wing from front to back, and preferably twice this length. Some notchback coupes and sedans may require an even taller wing mount if they have a steep rear window and short trunk. Naturally, the larger the wing, the more downforce and drag it creates.

The best location for mounting a wing depends on the design of the rear of your car. On a sedan or fastback, the wing usually goes at the very back of the trunk lid. This point usually has enough airflow for the wing to work effectively, except in a few cases where the trunk lid is short and the airflow is blocked by a steep rear window. On blunt-tailed cars, the wing goes on the roof in front of the hatch. Attempting to mount the wing below the rear window on such a car usually puts the wing in an area where there is no airflow at all, or worse, a spot where the air flows upward and will make the wing create lift.

Many aftermarket wings offer an adjustable angle. Raising the trailing edge will increase downforce and drag, up to a point. Increasing the angle too much can cause the wing to stall, a condition where the wing produces a tremendous amount of drag but virtually no downforce.

Wings sometimes come with additional features to improve their performance. End plates and other vertical plates sometimes appear on automotive wings. End plates are designed to reduce drag by controlling the way the air flows around the tip of the wing. Some wings use an additional strip at the trailing edge called a Gurney flap to increase downforce. Rally wings have recently added vertical plates in the middle of the wing as well. These plates are designed to keep the air flowing in the right direction over the wing even if the car is sliding sideways around a corner. These might also make sense in drifting, but for a road race car, this is not worth the extra drag.

When installing a wing, be sure to attach it to a point that can support downforce. One extreme case is the gargantuan wing on the Dodge Charger Daytona and Plymouth Roadrunner Superbird. Designed for NASCAR racing, this wing could create up to 700 lb of downforce on a superspeedway. The wing was mounted outside the trunk lid, high enough so that the trunk lid could be opened without touching the wing. Braces inside the trunk connected the wing directly to the rear subframe. While this is an extreme case, mounting a wing to a flimsy part of the bodywork can spell disaster if the wing produces any downforce. Installing a functional wing on a carbon fiber trunk lid can be asking for trouble, unless the lid was specifically designed to support a wing.

There are several material choices for wings. Aluminum wings are quite popular for racers and are often quite functional. However, a fiberglass wing is often much easier to paint and integrate into the look of a street car. Carbon fiber wings are also available for those who like the look or want to save weight. Most composite wings also feature streamlined mounting struts for slightly less drag than the typical aluminum wing.

Wings typically require drilling a few holes to install. If you are installing one yourself, be very sure that the mounting holes are in the right place before drilling. If

The size of the wing on the Plymouth Roadrunner Superbird and its high location, out of the way of the air coming off the roof, let it create massive downforce.

you misplace the holes, or decide you do not want this wing after all, repairing the holes will typically require filling them with welded metal and repainting.

Spoilers

Spoilers are often confused with wings. Both sit on the back of the car and can create downforce if correctly designed. The most visible difference is that a spoiler is attached to the bodywork, while a wing is mounted on pedestals to separate it from the

The spoiler on this Camaro separates the airflow from the body at the end of the trunk.

bodywork. Spoilers range in size from the very tiny ridge on the top of the rear window on some hatchbacks to the units found on NASCAR Sprint Cup cars.

Spoilers may be designed to either reduce drag or create downforce, but seldom do both. Smaller spoilers that protrude only 1 inch or so from the bodywork can reduce drag if designed correctly. Larger ones, which may be 8 inches long, can generate downforce by pushing airflow upward.

The correct location for installing a spoiler is often the same as for wings. Blunt-tailed cars need the spoiler installed at the top of the rear window, while sedans, notchback coupes, and fastbacks use trunk-mounted spoilers. Some sedans and notchbacks also benefit from having a roof spoiler installed at the top of the rear window.

Installing a spoiler can range from simply sticking it in place with double-sided tape or screws to complicated bodywork to blend it in with the trunk lid and make the installation look seamless. The latter usually involves applying body filler to the base of the spoiler and sanding it until there is no apparent gap between the spoiler and the bodywork. If you want this sort of installation to look its best, it may be worth the money to have it done professionally at a body shop.

Front Bumpers, Airdams, and Splitters

Body kits were quite popular in the '90s, and some cars still have a decent selection available today. You may want to pick one that makes the sort of fashion statement you want for your car, but if you want a functional choice, there are several things to consider. If you have an older car and want to go road racing, you may even want to fabricate your own parts for your front end, as the right mods up front can reduce drag, add downforce, and help cool various things under your hood.

One of the most important considerations if you are replacing your front bumper is safety. Older cars used solid steel bumpers to absorb impacts, but on newer cars, the bumper is often a rubbery "skin" with a framework of impact-absorbing plastic underneath it. For safety, you will want a bumper that lets you keep the original reinforcement and replaces only the outer layer. For street use, you will also want to get a front bumper made from urethane or other flexible material instead of fiberglass. Fiberglass bumpers are very easy to crack on speed bumps or parking lot curbs. Also, if

This Monza has an airdam extending almost all the way to the salt.

you live in a state that requires front license plates, make sure the bumper has adequate provisions for mounting a plate. Some kits do not have any suitable mounting place, and others put the plate in spots that do not look particularly good.

One of the most important parts of the front end for performance is the airdam. This is a valence that extends from underneath the bumper to reduce the amount of air flowing under the car. Airdams are often molded into a one-piece assembly along with the bumper on modern cars, but older cars often require an add-on part. Preventing air from flowing under the car not only reduces drag (the undercarriage typically is rather rough), but also creates a partial vacuum under the car to create downforce.

Airdams often feature several openings. You will want to make sure that these are positioned so that they can direct air to the front brakes, radiator, intercooler, or other cooling systems. If necessary, you may have to add some sort of ductwork to make sure air coming in through an opening does not end up flowing under the car. Unnecessary openings in an airdam will diminish its effectiveness. Some road racing cars use almost no openings in their airdams at all.

Adding a splitter can maximize the effectiveness of an airdam. A splitter is a flat piece at the bottom of an airdam that makes it difficult for air to flow down the airdam and under the car. Splitters are sometimes molded into the airdam, but are often made as a separate part, using plastic, carbon fiber, or sometimes even plywood. Since splitters are the lowest part of a front end and stick forward, they are the most vulnerable to damage. Making the splitter a separate part reduces the odds that hitting anything with the splitter might also crack the airdam.

Some cars also add small winglets to the side of the front bumper called canards. These are much like miniature versions of rear wings and create a small amount of downforce at the front end. If the canards do not have good support, they can flex the front bumper, possibly cracking the paint.

Installing a front bumper is much like installing a spoiler. Some can bolt on as a direct replacement for the original bumper. Others, particularly the less expensive ones, require the services of a body shop to make them fit correctly.

Ground Effects

Side skirts are designed to continue the work of the airdam by restricting air from flowing in from the sides to the low-pressure area created by the airdam. Having the side skirts reach lower increases their effectiveness, but also makes them more vulnerable to road damage. Other than that, there is not much more to selecting side skirts than to choose a set that matches the aesthetics of your car well, or one that complies with the rule book, if you're going racing.

When using ground effects, the rear bumper should be designed to let air flow out from under the car. Most cars have a partial vacuum behind the trunk lid, and ground effects can take advantage of that to create a partial vacuum under the car. Rear bumpers should provide plenty of space, either in the form of cutouts or gratings, to allow for air to flow out. Of course, if you have a low airdam and side skirts, sometimes using the original rear bumper (which won't reach down as far as the skirts or front bumper) can accomplish the same thing as an aftermarket one with mesh openings. Like with front bumpers, it is a good idea to get a rear bumper made from polyurethane that retains the original reinforcement.

One related trick is to create downforce by giving the car a raked stance. Setting the front of the car an inch lower than the rear can be sufficient. The air is forced into a smaller opening at the front of the car, and then the increasing area between the undercarriage and the road helps pull the air along and reduce pressure under the car. The flip side is that a car that sits higher in the front than in the back can have serious stability problems.

One other under-car improvement is a belly pan. A belly pan is a flat piece of metal covering all the decidedly nonstreamlined parts underneath: the exhaust, drive shafts, fuel tank, and the like. Most belly pans are custom made. One popular trick is to extend the splitter back to the engine and front wheels. Further sections can be added by attaching sheet metal with screws or rivets. Just be sure to leave enough clearance for such things as suspension and the driveshaft to move. The biggest trade-offs are that belly pans add weight and get in the way of repairs.

Truck Bed Covers

The bed and tailgate of a pickup truck can create a significant amount of drag. Some people who drag race pickups report that they can pick up noticeable gains by running with the tailgate down, but many newer pickups have less drag with the tailgate up. In either case, removing the tailgate and replacing it with a net is likely to create more drag, as flapping netting will create a lot of turbulence. Nearly all trucks become more aerodynamic if the hole in the bed is covered, however. A rigid fiberglass cover may perform slightly better than a fabric cover, but both will cut down on the drag. Many of these fiberglass covers also feature locks to allow you to keep your cargo safe.

Scoops

There are many parts of a car that need fresh, cold air. The engine often benefits from picking up cold outside air. Brakes last longer with air for cooling. Radiators and other equipment also need to have a ready source of airflow. The way to provide this is through properly placed scoops and inlets. An inlet is simply an opening in the bodywork, while scoops stick up to catch incoming air. Either one can create a "ram air" effect at high speeds, where the speed causes the pressure of the incoming air to rise. This can add a

The large inlet on the hood of this Ferrari gets the air in without creating too much drag. (*Photo courtesy DIYAutoTune.com.*)

This Camaro sports a massive forward-facing hood scoop. (*Photo courtesy DIYAutoTune.com.*)

few horsepower if ducted down the intake of your engine, but you must take care to prevent the pressure from leaking out. Ram air is most likely to be effective at speeds of 100 mph or so, but the cold air is still beneficial no matter how fast you are going.

Inlets often get the job done with less drag, but only work well when the airflow cooperates. Locating an inlet on the airdam or other vertical, forward-facing surface is one option. For flat surfaces, there is the NACA duct, designed by NASA's predecessor the National Advisory Committee for Aeronautics to draw in air that is flowing over a nearly flat surface. It does little good to mount an inlet in a spot where the airflow has separated from the car's bodywork. If the front end has sharp edges at the sides or the hood, the airflow will often separate at these edges.

Scoops are more blatant. A well-designed scoop will create its own high-pressure area to draw in air. These are very popular on the hoods of drag cars, where they provide both fresh air to the engine and extra clearance for extreme engine mods.

An unusual sort of scoop is the cowl induction hood. This tall hood is popular with older muscle cars, and has a large bulge in the hood with an opening at the base of the windshield. The reason this seemingly backwards scoop works is that a high-pressure area forms at the base of the windshield. Air will flow away from the high pressure in any direction available except forward, whether over the windshield, into vents in this area, or down the cowl hood scoop. It is possible on some cars to build a "stealth cowl

A cowl hood on a Mustang. (*Photo courtesy DIYAutoTune.com.*)

air" system by using the existing vents in this location, which often provide air for the car's interior.

Extractors and Louvers

Any air you draw into the car must be carried out. Normally, air in the engine compartment or fenders will escape out under the car. Unfortunately, this also results in quite a bit of pressure under the hood, making for more front-end lift. Adding extra ways for air to escape can reduce lift, as well as help cool the engine by allowing more air to flow through the radiator. Two means of doing this are extractors, which are essentially scoops turned inside out, and louvers cut into the bodywork.

Unlike inlets, these work at their best in an area where the airflow is separated from the body. In some cases, letting air out into an area where airflow is separated can reduce drag, which is why you sometimes see front bumpers with small louvers in front of the front wheels.

Fender louvers are a popular mod, although these would require ducts and holes in the inner fender to let out any air from the engine compartment. However, if the louvers can draw air out of the wheelwell, this can sometimes reduce front-end lift. The Charger Daytona used a set of extractors mounted on the tops of the front fenders to let out air and keep the front end planted at NASCAR speeds.

Fitting Oversized Wheels

Sometimes, you want to add larger wheels and tires than the stock bodywork allows. This may make you want to give up and settle for less rubber, but this problem can be overcome with enough money and determination. If you aren't getting too wild, you may need just the determination.

The cheapest way to fit wider tires is to roll the inner fenders. At its mildest, this simply takes the inner metal lip on the fender and folds it outward, preventing it from cutting into the tire. A little more force can push the sheet metal ever so slightly outward to give you a little more clearance. Minor fender rolling can be done at home with a baseball bat by taking the bat, placing it between the fender and the tire, prying outward, and rolling the bat back and forth. This can work extremely well if you just need to fold the inner lip out of the way. If you want to stretch the fender more, however, it can be difficult to get good-looking results by customizing your car with a baseball bat. A professional customizer can use hydraulic equipment or hammers to stretch a steel fender outward by several inches.

Sometimes the aftermarket can come to the rescue with fender flares or a wide body kit. Fender flares are usually kits designed for simple installation and may attach to the existing fenders to move them outward slightly. Note that not all fender flares actually give you more clearance; some just make the fenders look bigger. The ones that do give you more clearance need some sheet metal cutting to install. A wide body kit generally refers to a more complicated set of parts. Wide body kits replace the original fenders with new bodywork made from metal or composite materials. Often, these are patch panels that only replace part of a panel, requiring cutting out the sheet metal being replaced. Installing patch panels generally means having the services of a professional body shop, but a well-installed wide body kit can often look better than an easily installed set of aftermarket fender flares.

These options are useful if you need more clearance to the outside. There are also often ways to get more clearance from the inside, at least at the rear. Usually, the first limitation is the suspension itself. On cars with leaf springs, it is possible to relocate the springs inward by welding on new mounting points and cutting off the originals. If necessary, the axle can also be professionally narrowed. Narrowing an independent rear suspension is extremely difficult, but there are a few suspension shops that are up to the task.

The other limitation is the inner wheelwells. This can be overcome by tubbing, which involves cutting out the original wheelwells and welding in new "tubs" to accommodate extremely large tires. A less extreme version is mini-tubbing. This means cutting a section out of the original wheelwells and widening them by welding in a strip of sheet metal.

Tubbing and mini-tubbing are popular with drag racers. They allow wide slicks without adding extra drag from flared fenders. The downside is that a narrowed suspension often has less roll resistance. Moving the tires out and flaring the fenders is usually better from a handling perspective, but will give you more drag.

Paint

Painting a car may not seem as daunting as swapping an engine or building your own engine control unit (ECU), but a quality do-it-yourself paint job can be more difficult than either one for a beginner. To make a top-quality paint job can require over a thousand dollars' worth of equipment and a considerable amount of experience. However, a novice who wants any body work done can benefit from knowing about the different sorts of paint and the preparation needed to paint a car.

Good preparation is even more important to getting a good paint job than good paint. All body damage should be repaired before painting. Rust will soon show through, even if chemically treated. It should be removed completely by grinding, and if necessary, welding in new patches of sheet metal. Low spots or other deformities that seem nearly invisible when a car is covered in flat gray primer can suddenly jump out at you when coated with glossy paint. It is not too difficult to paint over original paint in good condition, but if your paint is starting to show cracks or other signs that it is falling apart, it will take the new paint with it. If a car has several layers of paint already on top of the original, these extra layers may result in a thick layer of paint that is prone to cracking. Such a car should be sanded down to its original paint, and if the original paint is bad, sanded down to bare metal before painting.

Cleanliness is of prime importance in painting. Any oil or other contamination on the surface can cause problems for the paint. Although the paint shop should do their own washing, it doesn't hurt to wash a car thoroughly before bringing it to the paint shop. If possible, you may want to take a look inside the shop's spray booth to make sure that they keep their painting area clean. Dust from previous paint jobs can wind up in your paint if the shop is not careful.

Paint comes in several varieties. The paint on most new cars is polyurethane based, using a very similar chemical to the one used to make suspension bushings and bumpers. Polyurethane is a very durable, glossy paint whose main drawback is the expensive safety gear required to spray it. Acrylic enamel may not be quite as shiny, but it is closer to the sort of paint used from the factory in the muscle car era. Using enamel on these cars can make for a more authentic-looking finish. Another sort of paint, acrylic lacquer, is seldom used except on cars from the '50s and earlier. Recently, a few paint companies have come out with water-based paints suitable for automotive use. Currently these require a clear polyurethane top coat. However, these provide good results without the hazardous solvents used in other paints. If and when it becomes possible to paint a car entirely with water-based paint, it may again be practical to paint a car in your driveway without annoying environmentalists (although many states will still let you paint a car in your driveway if it's your own car and you're not painting for money). Any sort of paint will first require a coat of an appropriate primer unless it is going on over existing paint.

While paint shops can mix nearly any color, if you want to change the color of your car, the most practical choice is to go with a color already used on production cars. This can make matching the color much easier if your car is dented and needs any touch-up paint, as a body shop can look up the formula to match the color in their guide books.

There are several extra features that you can add to paint for different looks. One of the most popular additions is a clear coat. The clear coat is simply a layer of clear paint applied over a colored base coat. The clear coat can make paint look glossier as well as protecting the paint. The main downside (besides the expense of another coat of paint) is that a clear coat can be a little bit more difficult to repair. Kandy is similar to clear coat with translucent color mixed in. Kandy requires a colored base coat or colored primer, and the final color will show traces of what's under it. This paint requires a very skilled painter, and matching this paint correctly for touch-up work is nearly impossible. Metallic paints are more common than kandies. These use a small amount of metal flake added to the paint to give it a sparkling look.

Powder coating is also a process worth mentioning. This process is usually not practical for painting an entire car body unless you are starting with a completely disassembled car stripped to bare metal, but it can be used for engine blocks, suspension parts, and the like. Essentially, it is painting without solvents. Powder coating uses a plastic powder that sticks to the parts by static cling and is then baked on in an oven until it melts and sticks to the part. The result is a rugged finish, although it often doesn't look quite as good as a fresh coat of paint.

Graphics and Decals

While a solid color can be quite eye catching, many enthusiasts want to spice things up further with graphics. These may be painted on or added with vinyl decals. Often, these two approaches fit different sorts of graphics.

Painted-on graphics are usually done with an airbrush, although some are done with smaller paint guns, or miniature rollers in the case of pinstripes. In the case of scallops or other simple designs, you can even mask off the car with tape and newspaper, and apply them with a spray can. The most common example of painted-on graphics is probably the flame job, which needs no introduction. Scallops are another painted-on graphic, which usually start with a solid color around one end that tapers to a set of tapered stripes following major body lines. Pinstripes are very thin stripes that can be as simple as a single pair of stripes running the length of the car or as complicated as the ornate Van Dutch patterns.

Vinyl decals are used both for small sponsor decals and large graphics on the side of the car. Laying out the graphics with vinyl often allows for designing the graphics by computer and even seeing how the finished result will look before they are applied. The

Old-fashioned, painted-on graphics.

downside is that vinyl may not be able to handle some sorts of blended colors as well as a skilled painter with an airbrush, and effects like metallic paint are nearly impossible. Local sign shops can often make custom vinyl graphics if you can supply the image you want in a suitable computer file.

While complex paint is best applied by professionals, many vinyl graphics are easy to apply at home. Smaller ones can simply be applied using a plastic squeegee and carefully working out the bubbles. Larger stretches of vinyl often call for using a soapy water solution to wet down the area and stretching the graphics tightly to avoid air bubbles. After you are done, small bubbles can be removed by gently puncturing them with an X-Acto knife, taking care to avoid touching the underlying paint. Then heat the bubble with a hair drier gently to make the vinyl shrink. Or you can have a graphics shop or window tint installer install the vinyl, and save yourself the hassle. Installing large graphics without wrinkling or bubbles can be tricky.

Wraps are a new trend that takes vinyl to a new level, covering an entire car in printed vinyl decals. These are cheaper than custom paint over a whole car, and let you lay out complex art in advance. The downside is that a wrap may not look quite as good close up as top-quality paint.

Sometimes, you may want graphics that are easily removed or changed. For example, many people who autocross their street cars like to remain inconspicuous when driving to and from events. The solution is to have graphics printed on flexible magnetic material, much like oversized refrigerator magnets. As with vinyl, local sign places can often make magnetic graphics for you, as well as vinyl that holds on by static for use on fiberglass or aluminum cars. If you are on a budget, you can even buy magnetic material where they sell office supplies and print it with a laser printer or inkjet.

Lighting

Lighting upgrades are also popular, sometimes for looks, and sometimes because it seems the original headlights do not put out sufficient light. Headlights can be upgraded or supplemented in a variety of ways. Taillight mods may not have any functional benefits, but they have been popular since the days of customizing Model Ts.

When it comes to visibility, there are three factors to consider. The first is the total brightness. The second is the color. Blue light is more likely to create glare for oncoming drivers or scatter in fog, while yellow or amber lights often carry further in such weather than other colors. The third characteristic is the beam pattern. The way the light leaves a bulb is not distributed evenly. Most headlights have a cut-off, a level above which they put out very little light to avoid blinding oncoming drivers. While some light at the very front of a car is useful, having too much light right in front of the car can make further objects look dimmer. Lights must strike a good balance between illuminating near and far objects. Beam patterns are also different depending on which side of the road the car is intended to drive on, so if you import lights (or whole cars) intended for use in a different country, you may wind up with lights that put too much light in the oncoming lane and not enough on the side of the road.

Automotive lights come in two basic types. Incandescent lights, like Edison's first light bulb, work by heating up a metal filament in a gas-filled chamber until it glows. The efficiency of these bulbs can be improved by adding different gases to the chamber, as in halogen lights that are partially filled with halogen gases. Most are slightly yellowish in color.

High-intensity discharge (HID) lights work by creating an electrical arc between two electrodes. These require a high-voltage power supply, but are much more efficient. This increased efficiency allows for brighter lights. These are sometimes called xenon lights, although the term xenon is also used to describe halogen lights that have had a little xenon added to the gas mixture, or sometimes even lights that have been tinted to imitate the color of HID lights. HID lights are normally pure white with a little extra blue mixed in.

If you have old-fashioned sealed-beam headlights, there are several upgrades on the market that easily swap in. One possible upgrade is to swap in a set of European spec lights, known as E-code headlights. These lights are not always officially street legal, but they can put out more light where you need it. E-code replacement lights often replace the one-piece bulb assemblies with a housing featuring a removable bulb. If you do not want to go this far, a high-end halogen light can offer an improvement over the cheaper designs.

A number of dubious bulb "upgrades" are on the market for U.S.-spec headlights with replaceable bulbs. The most common was a regular bulb with a coating meant to make them look bluish, supposedly to resemble HID lights. This coating works by filtering out several of the colors. The end result, not surprisingly, is a dimmer bulb. Thankfully, these seem to be on the way out. One other questionable choice is to install a higher wattage bulb without checking if your reflector or lens can handle it. This does put out more light, but these bulbs run considerably hotter than the original bulbs and can melt plastic reflectors or lenses. A high-wattage bulb can work safely, however, if your headlight assembly uses a glass lens and a metal reflector.

Sometimes, you may see kits for installing HID lights on cars that originally came with halogen lights. There is a right way and a wrong way to convert halogen lights to HID. Some kits simply replace the bulb with a special HID module. The trouble with this approach is that the reflector and lens are designed to work with the original bulb, and the glowing area of an HID bulb has a different shape. Using the original lens and reflector usually ends up with a pattern about halfway between a high beam and a low beam. The correct way to convert to HID lights is to replace the entire headlight assembly: bulb, reflector and all. A common method is to take small factory HID lights, like the ones off an Acura TSX, and stick the whole assembly behind the original lens. Use parts that are designed to work together.

If you need more lighting and cannot find an easy way to upgrade your headlights, a good option is to use auxiliary lights. These lights come in two basic types: driving lights and fog lights. Each one has its own use. Fog lights are meant for driving very slowly through thick fog, and should be installed so they can be switched on independently of the headlights. A properly designed fog light will illuminate the area immediately in front of the car, letting you see the lane lines. Good fog lights can be yellow, white, or amber. Driving lights are designed to light up faraway objects and help out the headlights, and are usually white. It's a good idea to wire your driving lights so they come on when you turn on the high beams. The amount of light put out by the bulbs is usually measured in lumens and marked on the box, allowing you to compare the output of different driving lights. Some on the market are weaker than normal headlights, so check carefully.

Headlights, fog lights, and driving lights all need to be aimed correctly. Most of them include provisions for adjustment built into their housing. Besides adjusting them when you first install new lights, you should aim your headlights any time you change the suspension's ride height. It's easy to forget about your headlights when you lower the front for a nice raked stance, but this will make your headlights point down too.

One taillight trend that comes and goes is to take taillights from one model car and graft them onto another. This might require cutting new holes in the bodywork or filling in the area occupied by the old taillights with fiberglass, Bondo, or sheet metal. If you can find a set of taillights that works well with the lines of your car, the effect can be subtle but impressive. Grafting on production taillights also helps assure that you are installing a set of taillights already designed to meet government requirements.

Recently, the aftermarket has started mass producing custom taillights that install easily in place of the original assemblies. This can be much easier than dealing with the bodywork involved with taillight swaps. If you are looking at a set of aftermarket taillights, make sure to get a set that meets all regulatory requirements. Police can and do ticket cars for illegal taillights. In particular, you will want a red reflector visible from the rear and side. The lights absolutely must be red, except that turn signals may be amber, and, of course, reverse lights are white.

Gearhead Glossary

The way car people talk can confuse an outsider. Some of the terms are straightforward words for parts or racing terms, while others are a sort of slang that reflects car culture.

Aftermarket   Refers to parts intended for installation on a car after it is purchased from the dealership, usually designed to differ from the parts on the car when it left the factory. As a noun, the market for such parts, and the companies that supply them.

AN fittings   Originally developed for the military (AN means Army–Navy), this standard of fitting is often used on race cars. Most AN fittings work with braided hoses.

Anti-roll bar   A flexible bar that ties the left and right sides of the suspension together to reduce body roll.

Autocross   A racing event where cars run one at a time against the clock around a course laid out with cones in a large parking lot, an extremely tight road course, or any other low-speed racecourse with sharp corners.

Barrel   Many carburetors contain several barrels. These are the basic "building blocks" of a carburetor. Each barrel has its own separate throttle and discharge jets to release fuel into the incoming air. Some carburetors open all the throttles at the same time, but many open the throttles in one set of barrels before opening the others.

Base model   The least expensive version of a particular car. Base models frequently have fewer luxury items, less powerful engines, and softer suspensions than the more expensive versions.

Billet   A billet part is made by taking a solid piece of metal (also called a billet) and cutting it to shape. Billet parts are often found on show cars, since this technique is the easiest one to use for making a single part.

Blow-off valve   A valve located between a supercharger or turbocharger and the throttle. This is designed to vent any pressure spikes caused by suddenly closing the throttle.

Blower   Slang for a supercharger.

Boost   On an engine running a turbocharger or supercharger, boost is the pressure level in the intake system compared to the outside air pressure. If the air pressure outside the engine is 14.7 psi and the absolute pressure of the air being fed into the engine's intake is 21.7 psi, the engine is said to be running 7 psi of boost.

Bore   The diameter of the cylinders of an engine. Also, the process that cuts the bore to a precise size.

Bracket racing   A type of drag racing where one car is given a head start based on predicted performance. Successful bracket racing requires the driver to be able to consistently run the same time in every race.

Braided hose   A braided hose features an inner plastic hose protected by an outer layer of braided metal. Braided hoses are stronger and stiffer than standard rubber hoses.

Burnout   The act of spinning a car's tires while the car is either stopped or moving very slowly.

Bushing   A rubber, plastic, or metal tube-shaped part used to allow one part to pivot around another while absorbing minor vibrations. Usually found in the suspension.

CAI   Cold air intake.

Caliper   A hydraulic device that squeezes a pair of brake pads against a disc brake. A "pair of calipers" can also mean a precision measuring device.

Camshaft   A spinning metal rod with several protruding lobes. The camshaft determines when the valves in the cylinder head open and close.

Capacitive discharge   A type of ignition amplifier that stores electrical energy in a capacitor.

Carb   A carburetor.

Carbon fiber   A material made by weaving threads of pure carbon into a fabric and encasing the fabric in glue. Carbon fiber is very lightweight but brittle.

Carburetor   A mechanical device combining a throttle control with a series of passages to mix fuel with the incoming air.

Cast   A cast part is made by pouring molten metal into a part-shaped hole in a block of metal, ceramic, or sand. Casting is often a very inexpensive way to make a large number of parts.

Cat-back   An exhaust kit for replacing all parts in the exhaust system downstream of the catalytic converter.

Catalytic converter   A device in the exhaust that reduces pollution by breaking down byproducts from incompletely burned fuel.

Christmas tree The arrangement of starting lights at a dragstrip. Each side has two yellow lights to indicate the car's position at the starting line, three amber lights to count down before the start, a green light to signal the start, and a red light to indicate an early start.

CNC machined Sometimes advertised on performance parts, CNC machining simply means that the machines that did a particular task operated under computer controls. This can be important if it calls attention to a part that is normally not machined, such as if a cylinder head has extra port work.

Coil-over An assembly that combines a shock absorber or strut with a coil spring. Usually features an adjustable spring mount, allowing the owner to change the car's ride height.

Control arm A part of the suspension, usually triangular, used to connect the chassis or car body to the spindle where the wheel is mounted.

Cowl induction hood A hood with a large scoop where the opening is at the bottom of the windshield.

Cross-drilling Drilling a pattern of holes in a part, usually the brake rotor. Usually done for cosmetic or weight reduction purposes, but this weakens the part.

CV joint A part that connects two sections of a shaft and allows the shaft to bend while turning. CV stands for constant velocity, because the shaft parts spin at the same speed, regardless of the angle.

Damper A device that is meant to absorb the vibrations of the car's springs. Dampers usually fall into two categories: struts and shock absorbers.

Displacement The total volume of the cylinders inside an engine.

DSM Diamond Star Motors, a joint venture between Chrysler and Mitsubishi that built the first- and second-generation Mitsubishi Eclipse as well as the Eagle Talon and Plymouth Laser.

Dynamometer A machine for measuring horsepower and torque.

Dyno See *Dynamometer*.

ECU Engine control unit. A computer that controls fuel injection and spark timing.

EFI Electronic fuel injection. A system that uses fuel injection controlled by a computer to determine how much fuel the engine needs.

Elapsed time The time needed to cover a race course, especially a dragstrip.

End link A small metal rod that connects an anti-roll bar to the suspension.

E.T.   See *Elapsed time.*

Fenderwell headers   Headers that cannot be installed without cutting the inner fenders.

Forged   A forged part is made by taking a solid (although sometimes very hot) piece of metal and squeezing it between two tools known as dies. The piece of metal takes the shape of the dies. A forged part is often stronger than a cast part.

Fuel injection   A system that mixes fuel with air drawn into the engine by spraying the fuel through a nozzle. The amount of fuel delivered is determined by how the valve opens and closes. Fuel injection is usually controlled by electronics, but can be controlled mechanically.

Glasspack   A muffler design with a single pipe at its center with holes or louvers directing gases into an outer chamber lined with fiberglass.

Header   An exhaust system component that runs an individual pipe from each cylinder and joins the pipes into a larger pipe.

Holeshot   In drag racing, gaining an advantage by leaving the line before the other driver.

Hybrid   A car where the original engine has been replaced by a more powerful one originally installed in a different model. This term is usually used with Hondas and other Asian imports. Also refers to cars driven by both a normal engine and an electric motor.

Ignition amplifier   An electrical box designed to take a weak signal from a sensor or ECU and amplify it to drive a coil to fire the spark plugs.

Injector   A valve that opens and closes rapidly to spray fuel into the air entering an engine.

Intercooler   A device resembling a radiator and designed to cool air exiting a turbocharger or supercharger.

JDM   Japanese domestic market. Cars or parts made in Japan and intended for Japanese use.

Jet   A small plug with a precision hole drilled in it, used in many carburetors. To rejet a carburetor means to retune it to give the engine a richer or leaner air/fuel mixture.

Lateral g   A measurement of a car's cornering ability. A car cornering at one lateral g applies a force equal to its own weight in cornering traction. Running in a circle 200 feet in diameter at 38.7 mph requires one lateral g of grip.

Limited slip differential   A device that enhances traction by preventing one wheel from standing still while the one on the opposite side of the car is turning.

Manifold   A component that air or liquid flows through that connects a group of similar objects, such as connecting one throttle body to four cylinders. Manifolds are often made of cast iron or aluminum and used in both the intake and exhaust system. Some antique cars also have manifolds in the cooling or oiling system.

Motor mount   Component used to hold the engine in place while absorbing vibrations. Usually contains a block or cylinder of rubber with mounting brackets.

Motor plate   A type of motor mount that is made from a solid metal plate and bolts to the front of the engine instead of the usual side locations.

OEM   Original equipment manufacturer. Technically this means the company that originally made the car or a part of the car. However, this is often used as an adjective to indicate that a part is as delivered from the factory. Ironically, several OEM companies also build aftermarket parts.

Plenum   A central chamber in an intake manifold that holds a large volume of air. The plenum is typically connected to individual runners leading to each intake port on the cylinder heads.

Polyurethane   An amazingly versatile plastic. Polyurethane is somewhat rubbery and can be used to make suspension bushings, bumpers, interior trim parts, gaskets, caulking, and even paint. Polyurethane is stiffer than most natural types of rubber.

Port   An opening in the cylinder head to let air in or exhaust out. As a verb, used to mean reshaping the ports on a cylinder head to increase power.

Posi   Short for Positraction, General Motors' trademark for a limited slip differential.

Positive displacement supercharger   A supercharger that produces a nearly constant amount of boost at full throttle, no matter what the engine speed.

PSI   Pounds per square inch, a measurement of pressure.

R-compound tire   A tire meeting at least the minimum requirements for street use, but designed primarily for road racing or autocross where rules require cars to run on street tires. R-compound tires generally have very little tread depth and use very soft rubber for their tread.

Rally   An event where cars are raced one at a time on gravel or dirt roads.

Resonator   A small exhaust component resembling a muffler but providing less restriction and less silencing. Often works on the same principle as a glasspack muffler.

Riceboy   Refers to someone who either attempts to make his car look fast without making a serious effort to increase the car's performance, or installs minor performance mods and severely overestimates how fast his car is. Common examples of touches a riceboy might add include stickers for parts he does not have, fake oversized brake rotors, emblems from a car with a more powerful engine, and the like. This term is usually applied to owners of Japanese cars or front-wheel-drive American cars, but owners of Mustangs with fake 5.0 emblems or similar posers would also qualify.

Rim   The outside part of a wheel where the tire is attached. Often used to refer to the entire wheel.

Shock absorber   A device for damping out the vibration of a car's springs. Shock absorbers can make a car handle in a more predictable manner and have an especially large effect at the very beginning and end of turns.

Shoehorn   To stuff an engine into a car that was not designed for a motor of that size.

Skid pad   A large paved circle used for measuring a car's cornering ability.

Sleeper   A car that has been planned to look slower than it actually is. Sleepers may be built by modifying a car that others would not expect to be fast, by attempting to hide modifications, or both.

Stock   Describes a car or parts as delivered from the factory.

Stroke   The distance a piston moves up and down in a cylinder. As a verb, it means to modify an engine to increase this distance. On rare occasions, you may see a destroked engine, one that has had this number decreased.

Strut   A shock absorber that is free to pivot at one end and bolted solidly to the suspension at the other.

Supercharger   An air pump that pushes air into an engine to increase power and torque.

Sway bar   Another name for an anti-roll bar.

Throttle   A valve that controls the speed of an engine by regulating the amount of air it can pull in.

Throttle body   An assembly on a fuel-injected engine containing the throttle and usually several sensors.

Torque   A kind of twisting force usually measured in pound-feet. Applying 50 pounds of force perpendicular to the end of a 1-foot-long wrench handle would put 50 pound-feet of torque on the bolt at the other end. So would putting one pound of force on a 50-foot-long wrench.

Torque converter   A fluid-based device that connects the engine to an automatic transmission.

Traction bar   A device used to prevent part of the suspension (usually a leaf spring) from flexing under hard acceleration. This can increase traction because many parts can flex in ways that pull the wheels upward.

Transbrake   A device that shifts an automatic transmission into two gears at once, often both forward and reverse. Since the transmission cannot turn at two different speeds, the car cannot move. Used in drag racing to allow the engine to fully load the torque converter with the car standing still before the car leaves the starting line.

Trap speed   The speed at which a car is moving at the finish line of a dragstrip. Trap speed can be used to estimate a car's power-to-weight ratio.

Tri-Y header   A header where the exhaust pipe splits into two pipes, and these two pipes each split into two pipes, giving four pipes that connect to the engine's exhaust ports.

Tunnel ram   A type of intake manifold that uses multiple carburetors above a rectangular open space (the "tunnel"). Each barrel of the carburetor is positioned above a nearly straight runner to each cylinder. Tunnel rams are usually considered race-only parts, but can be adapted for street use if you keep the carburetors small enough. Often requires cutting a hole in the hood or using a cowl induction hood.

Turbo lag   Turbocharged engines sometimes have a delay between when the throttle opens and when the turbo can "spool up" to provide maximum power.

Turbo muffler   A muffler loosely based on the design used on the '60s-era Chevrolet Corvair Turbo. Turbo mufflers are not necessarily designed for turbocharged cars.

Turbocharger   A centrifugal air compressor driven by a turbine mounted in an engine's exhaust system. Turbochargers force extra air into an engine to increase power and torque.

Urethane   See *Polyurethane*.

Variable displacement supercharger   A kind of supercharger that produces more boost as engine RPM increases.

Wastegate   A valve that directs exhaust away from the turbocharger to control boost levels.

Windage tray   A device that mounts in the engine oil pan and is designed to keep oil off the crankshaft.

Further Reading

There are many useful books out there for learning how to modify cars. The following books are ones that I have found very useful to beginners and apply to many different sorts of cars.

Bell, Corky. *Maximum Boost*. Cambridge, MA: Bentley Publishers, 1997.

Written in a conversational style, this guide to turbocharging contains a fair amount of practical tips and descriptions of turbo installations.

Cramer, Matt, and Jerry Hoffmann. *Performance Fuel Injection Systems*. New York, NY: HP Books, 2010.

My other book is a practical guide to electronic fuel injection.

Graham Bell, A. *Four Stroke Performance Tuning*. Somerset, UK: Haynes Publishing, 2012.

If you're looking for a very advanced and in-depth book on a wide variety of engines, this is a good one. He also has a companion book, *Forced Induction Performance Tuning*, covering turbocharging and supercharging. Neither is the most beginner-friendly, but both books offer more depth than a beginning book.

MacInnes, Hugh. *Turbochargers*. New York, NY: HP Books, 1984.

A solidly written and technically sound guide to selecting and installing turbochargers. Unfortunately, it contains very little information about new developments in fuel injection and electronics. Contains one of the few in-depth published descriptions of how to make a turbocharger work on a carbureted engine.

Puhn, Fred. *How to Make Your Car Handle*. New York, NY: HP Books, 1981.

Still considered the definitive guide to handling. While suitable for beginners, this book contains enough detail that it could provide an engineer with enough information to design a top-notch performance suspension. Contains a few bits of outdated information, such as about cross-drilled brake rotors and baking springs, but the formulas you will find in this book are timeless.

Vizard, David. *How to Build Horsepower*. North Branch, MN: S-A Design Books, 1990.

David Vizard has worked on a wide variety of cars, from American circle-track racing to European and Japanese imports. This book primarily deals with carbureted, naturally aspirated, pushrod American V8 engines. However, there are many useful tips in this book that can be applied to virtually any engine.

Watts, Henry A. *Secrets of Solo Racing*. Sunnyvale, CA: Loki Publishing, 1989.
A very useful guide to autocross and high-performance driving techniques.

Index